Zeikus

Mixed Culture Fermentations

Special Publications of the Society for General Microbiology

PUBLICATIONS OFFICER: COLIN RATLEDGE

1. Coryneform Bacteria,
   eds. I. J. Bousfield & A. G. Callely

2. Adhesion of Microorganisms to Surfaces,
   eds. D. C. Ellwood, J. Melling & P. Rutter

3. Microbial Polysaccharides and Polysaccharases,
   eds. R. C. W. Berkeley, G. W. Gooday & D. C. Ellwood

4. The Aerobic Endospore-forming Bacteria:
   Classification and Identification
   eds. R. C. W. Berkeley & M. Goodfellow

5. Mixed Culture Fermentations,
   eds. M. E. Bushell & J. H. Slater

Mixed Culture Fermentations

Edited by

M. E. BUSHELL

Department of Microbiology
University of Surrey
Guildford, Surrey, UK

and

J. H. SLATER

Department of Environmental Sciences
University of Warwick
Coventry, West Midlands, UK

1981

Published for the
Society for General Microbiology
by
ACADEMIC PRESS
A Subsidiary of Harcourt Brace Jovanovich, Publishers
London   New York   Toronto   Sydney   San Francisco

ACADEMIC PRESS INC. (LONDON) LTD.
24/28 Oval Road,
London NW1

United States Edition published by
ACADEMIC PRESS INC.
111 Fifth Avenue
New York, New York 10003

Copyright © 1981 by
SOCIETY FOR GENERAL MICROBIOLOGY

All Rights Reserved
No part of this book may be reproduced in any form by photostat, microfilm, or any other means, without written permission from the publishers

British Library Cataloguing in Publication Data
Mixed culture fermentations. – (Special
 publications of the Society for General
 Microbiology; 5)
 1. Enzymes – Industrial applications –
 Congresses
 2. Fermentation – Congresses
 I. Bushell, M. E. II. Slater, J. H.
 III. Series
 660.2'844'9   TP156.F4

ISBN 0-12-147480-1

LCCCN 81-68019

Printed in Great Britain by
Whitstable Litho Ltd.,
Whitstable, Kent

CONTRIBUTORS

BAZIN, M.J. *Microbiology Department, Queen Elizabeth College, Campden Hill Road, London W8 7AH, UK.*

BUSHELL, M.E. *Department of Microbiology, University of Surrey, Guildford, Surrey GU2 5XH, UK.*

DRIESSEN, F.M. *Netherlands Institute for Dairy Research, Ede, The Netherlands.*

HOBSON, P.N. *The Rowett Research Institute, Greenburn Road, Bucksburn, Aberdeen AB2 9SB, UK.*

KIDNEY, E. *Bass Ltd., High Street, Burton on Trent, Staffs DE14 1J2, UK.*

LEE, Y.K. *Microbiology Department, Queen Elizabeth College, Campden Hill Road, London W8 7AH, UK.*

SLATER, J.H. *Department of Environmental Sciences, University of Warwick, Coventry CV4 7AL, UK.*

SOMERVILLE, H.J. *Concave, Van Hogenhoucklaan 60, 2596TE, Den Haag, The Netherlands.*

WHITE, F.H. *Bass Ltd., High Street, Burton on Trent, Staffs DE14 1J2, UK.*

WOOD, B.J. *Department of Applied Microbiology, University of Strathclyde, 204 George Street, Glasgow G1 1XW, UK.*

PREFACE

The presentation of a symposium on mixed culture fermentations at Queen Elizabeth College, London (December 1980) coincided with a renewed interest in and awareness of many aspects of biotechnology. This volume consists of reviews based on papers delivered at the Symposium, which was organized by the Fermentation Group of the SGM.

The modern fermentation industry has evolved largely from the development of technology capable of maintaining large scale monocultures. This includes processes for vitamins, amino and other organic acids and antibiotics. Publications in 1976 describing the Shell process for the production of edible protein from methane-utilizing microbial communities were, therefore, regarded as revolutionary in concept. Unfortunately, it was not possible to obtain a further contribution on this process for the Symposium but many other applications of mixed culture fermentations to biotechnology are presented here.

The first two chapters attempt conceptual and kinetic analyses of the interactions which occur in mixed cultures. The remainder of the book is devoted to descriptions of waste treatment processes and examples of mixed culture usage in the manufacturing industries. The involvement of microbial communities in waste disposal is such a wide ranging topic that we have differentiated between the principles involved in aerobic and anaerobic processes.

Developments in the fermentations for yogurt and beer manufacture are described, followed by a speculative account of the involvement of the yeast-lactobacillus interactions in the production of a number of "traditional" fermented foods and beverages.

Finally an aspect of the programme, under development at Queen Elizabeth College, for the microbial assimilation of solar energy is described.

This book has been edited by myself and Dr.J.H. Slater (University of Warwick) who also wrote the first Chapter, which introduces the subject.

M.E. Bushell
December 1980
University of Surrey.

CONTENTS

Contributors	v
Preface	vii

1 Mixed Cultures and Microbial Communities
 J.H. Slater

1. Introduction	1
2. Types of Microbial Community	3
3. Features and Problems Associated with Microbial Communities	15
4. Conclusion	20

2 Mixed Culture Kinetics
 M.J. Bazin

1. Introduction	25
2. Definitions	25
3. Competition	27
4. Commensalism and Mutualism	34
5. Predation and Parasitism	36
6. Multispecies Systems	44
7. Conclusion	49

3 Microbial Pathways and Interactions in the Anaerobic Treatment Process
 P.N. Hobson

1. Introduction	53
2. Digester Feedstocks and Design	54
3. The Overall Metabolic Picture	59
4. The "Passenger" Bacteria	60
5. The Breakdown of Polymeric Carbohydrates	62
6. Degradation and Utilization of Nitrogen Compounds	66
7. Some Miscellaneous Reactions	66
8. Lipid Metabolism: Production of Methane	68
9. General Observations	73
10. Conclusion	75

4 Mixed Cultures in Aerobic Waste Treatment
H.J. Somerville

1. Introduction	81
2. Microbial Degradation of Collected Wastes	83
3. Physiological Aspects	86
4. Transients in Substrate Concentrations and Other Operating Parameters	89
5. Separation of Biomass	91
6. Possible Developments	92

5 Protocooperation of Yogurt Bacteria in Continuous Culture
F.M. Driessen

1. Introduction	99
2. Batch Cultures	99
3. Continuous Cultures	104

6 Yeast-Bacterium Interactions in the Brewing Industry
F.H. White and E. Kidney

1. Introduction	121
2. Aggregation Amongst Microorganisms	121
3. Occurrence of Bacterial Contamination	122
4. Yeast-Bacterium Aggregations	122
5. Effect of Medium Composition	123
6. Turbidometric Assay	126
7. Role of Calcium	130
8. Cell Wall Protein Composition	130
9. Conclusion	133

7 The Yeast/Lactobacillus Interaction; A Study in Stability
B.J.B. Wood

1. Introduction	137
2. Beverages	137
3. Foods	141
4. Flavourings	143
5. Animal Feed	144
6. Milk-based Foods	145
7. Antibiotic Activities of Lactic Acid Bacteria	146
8. Conclusion	147

8 The Use of Algal-Bacterial Mixed Cultures in the Photosynthetic Production of Biomass
Y.-K. Lee

1. Introduction	151
2. Selection and Isolation of Mixed Cultures	152

3. Algal-Bacterial Interactions in Mixed
     Cultures   156
  4. Energy and Mass Balance of Mixed Cultures   160
  5. Stability of Mixed Cultures   163
  6. Prospects of Mixed Cultures   166
  7. General Conclusion   166

Subject Index   173

Chapter 1

MIXED CULTURES AND MICROBIAL COMMUNITIES

J. HOWARD SLATER

Department of Environmental Sciences, University of Warwick, Coventry, Warwickshire, UK

1. Introduction

For most microbiological studies and processes mixed cultures are considered to be the antithesis of good experimental techniques and practice. Microbiology students are rightly inculcated with the principles of pure culture techniques since much of our understanding of the properties and behaviour of microbes stems from work with axenic cultures. This is a necessary approach and tradition which, of course, remains as the cornerstone of experimental microbiology. It is due to the unique insight of Robert Koch who, one hundred years ago, published the first account of the method which reliably enabled him to isolate and maintain microbial cultures containing a single species. The method depended on microbial growth on the surface of a suitable, solid growth medium, the formation of microbial colonies starting from a single cell and, in principle and practice, remains unchanged to this day [Koch, 1881]. The paper was entitled "Methods for the study of pathogenic organisms" and fittingly started on page one of volume one of the "Reports of the Kaiser's Health Office". In the introduction to this seminal report Koch said that he did "not believe it is too much to say that the most significant point in all studies on infectious diseases is the use of pure cultures" and in writing this drew attention to the major problems that microbiologists then faced: namely, the inability to resolve microorganisms into species and correlate different species with different diseases and processes. Indeed at that time many bacteriologists thought that there was only one type of microbe and that different diseases were manifestations of different properties of the same organism.

The fact that the nineteenth century bacteriologists strove so hard to develop a reliable isolation and pure culture technique, highlights a truism which is greatly underestimated because of our insistence upon the use of axenic cultures: that is, pure culture growth systems are highly unrepresentative of almost all the habitats which support the growth of microorganisms. Most natural habitats contain a wide diversity of microorganisms, even those habitats which, for our convenience and with our prejudicies, are defined as extreme. For example, the Dead Sea which has a salinity which is at least 10 times greater than that of the oceans, has been shown to support the growth of at least 10 bacterial species [Brock, 1969] and this number is several orders of magnitude lower than the species composition of most so-called normal habitats. The structural and functional diversity of microorganisms obtained from the same habitat is due, in part, to the heterogeneity of the habitat and reflects the fact that specific microbes have evolved to occupy successfully that characteristic niche. Nevertheless, microbiologists, and microbial ecologists in particular, have been aware for a long time that the "growth range" of different microorganisms overlap, despite the spatial and temporal properties of a particular habitat. This is a situation which can and does lead to a variety of basic types of interaction between different species [Slater and Bull, 1978]. It is our argument [Slater, 1978; Slater and Lovatt, 1981; Bull, 1980] that there may be important principles and properties of mixed microbial cultures which may have been overlooked and neglected in the past. This is clearly of considerable importance to microbial ecologists studying the capabilities of microbial communities in their natural habitats or in laboratory-based systems which simulate various features of the natural environment. But the properties of mixed culture growth may be of potential interest to the microbiological processing or biotechnological industries and these may not yet have been fully exploited. At the outset it has to be recognized that most high technology processing industries (e.g., antibiotic production, primary and secondary metabolite production and others) rely on pure culture systems, for which there may be no need for or advantage to mixed cultures. In others, notably the food and, to some extent, the beverage industries, mixed culture systems are necessary and traditionally used [see Driessen, White and Kidney, and Wood, this volume]. In most of these cases the mechanisms involved with mixed culture growth remain obscure.

In this chapter I shall review briefly some of the principles of interacting mixed cultures although the definitions of the various basic types of interactions are dealt with elsewhere [Bungay and Bungay, 1968; Meers, 1973; Slater and Bull, 1978; Kuenen and Harder, 1981; see Bazin this volume]. In particular this chapter seeks to outline some of the difficulties, limitations and consequences of mixed cultures and microbial communities.

2. Types of Microbial Community

There is a growing literature describing many categories of interacting assemblages of microorganisms, although one of the present difficulties is to know exactly how significant some of these associations may be. Furthermore there is considerable confusion over the terminology used to describe the interactions and communities.

In general terms there is a spectrum of microbial communities ranging from rather tenuous, "loose" associations to "tight" associations, a few of which appear to be obligatory. In one sense any environment which supports the growth of more than one population constitutes a community and, to a lesser or greater extent and at different times and under different conditions, the component populations inevitably interact. This may not occur at all times but may, for example, be restricted to periods when a particular nutrient is present at limiting concentrations and affects the rate of growth of more than one population. Although such circumstances may be particularly important from an ecological viewpoint, there is little of detailed significance in the species composition. Many of these loose associations depend on non-specific commensal relationships and, in many cases, simply describe the flow of carbon, energy and other growth requirements between different populations.

The communities at the other end of the range are, in the present context, more interesting, exhibiting properties and characteristics which may not be apparent with any other combination of microorganisms or may not be expressed by the component populations existing alone or both. For example, the complete mineralization of a compound may require the sequential metabolism of two or more organisms with the component populations alone being unable to transform the compound since, separately, they do not possess the complete genetic complement to code for the whole biodegradative pathway [Slater and Godwin, 1980]. These tight associations will be referred to as microbial communities, although elsewhere they are also known as consortia [Whittenbury, 1978], syntrophic associations [Pfennig, 1978] or synergistic associations as well as simply mixed cultures or interacting assemblages.

It is now possible to suggest a simple classification of the various microbial communities which have been isolated [Slater, 1978; Slater and Lovatt, 1981] (Table 1). This classification has been compiled on the basis of the approach various workers have taken to the study of microbial communities and one of the major difficulties at the moment is that in most cases the analysis of these communities is often incomplete. Furthermore, with a more detailed understanding of the

TABLE 1

Classes of Microbial Communities [*after Slater and Lovatt, 1981*]

---

1. Structure due to the provision of specific nutrients between different members of the community.
2. Structure due to the removal of metabolic products which may be inhibitory to the producing member of the community, including hydrogen transfer communities.
3. Structure and stability due to interactions which may result in the modification of individual population growth parameters resulting in a more competitive or efficient community (compared with component populations).
4. Structure due to the effect of a concerted, combined metabolic capability, not expressed by the individual populations acting alone.
5. Structure due to a cometabolic stage.
6. Structure due to the transfer of hydrogen ions.
7. Structure is the result of the presence of more than one primary substrate utilizer — in many cases the nature of the interactions are unknown.

---

exact nature of interactions, particularly with categories 3 and 7 (Table 1), some microbial communities may be placed in other classes. For example, some stable communities have been examined with respect to some basic growth parameters including the maximum specific growth rate, saturation constant, inhibition constants and others [Osman *et al.*, 1976; Bull and Brown, 1979]. In the case of a three-membered orcinol community [Osman *et al.*, 1976] the nature of the interaction between the primary orcinol utilizer and its two associated non-orcinol utilizers was not determined. However, it is probable that the three-membered community's effectiveness, in growth kinetic terms, was due to the removal of inhibitory products generated as a result of the metabolism of the primary pseudomonad. Thus this community may simply be another class 2 type.

2.1. *Class 1 Microbial Communities*

The class 1 communities are widely distributed, readily isolated and most of those studied so far involve either commensal or mutualistic relationships dependant on the provision or requirement of growth factors or amino acids. This type of mixed culture is sometimes said to exhibit syntrophism. A number of examples of this category of community are shown in Table 2 and some of the reasons why these communities seem to occur so readily are discussed in Section 3.2.

2.2. Class 2 Microbial Communities

Class 2 microbial communities are those where excreted materials, which may be toxic and growth inhibitory to the producing organism, are consumed by associated members of the community. A number of these communities have been well characterized [Wilkinson et al., 1974; Cremieux et al., 1977] when methane or methanol are used as the primary carbon and energy sources (Fig. 1).

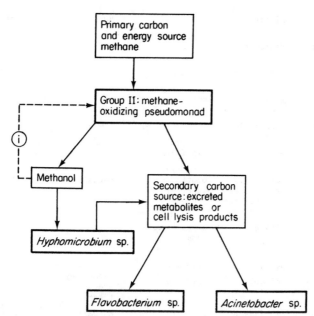

Fig. 1. A four-membered methane-utilizing microbial community illustrating the role of one population in removing an inhibitory metabolite produced by another organism (Wilkinson et al. [1974] after Slater [1978]).

The effectiveness of these communities was well illustrated by the four-membered community containing the *Hyphomicrobium* species, present to metabolize excreted methanol. If the mixed culture was stressed by the addition of extra methanol, methane oxidation by the primary pseudomonad ceased and there was an increase in the proportion of *Hyphomicrobium* sp. cells in the mixture: under the conditions examined the increase was from 4% to 25%. This ensured the rapid reduction of the methanol concentration, a resumption of methane oxidation and a gradual restoration of the original community composition. Clearly this community was more stable than a

TABLE 2

Class 1 Microbial Communities with Specific Nutrient Requirements

Organism	Compound product	Compound requirement	Type of interaction	Reference
Bacterium 3C1	—	Vitamin B_{12}	Commensalism	Jensen [1957]
Streptomyces spp.	Vitamin B_{12}	—		
Streptococcus faecalis	biotin	folic acid	Mutualism	Nurmikko [1954]
Lactobacillus plantarum	folic acid	biotin		
Streptoccocus faecalis	phenylamine	folic acid	Mutualism	Nurmikko [1956]
Lactobacillus arabinosus	folic acid	phenylamine		
Saccharomyces cerevisiae	riboflavin	—	Commensalism	Megee et al. [1972]
Lactobacillus casei	—	riboflavin		
Nocardia sp.	cell lysis products or cyclohexane catabolites	biotin (+ other growth factors?)	Mutualism	Stirling et al. [1976]
Pseudomonas sp.	biotin	organic carbon from *Nocardia* sp.		
Mucor	pyrimidine	thiamine	Mutualism — producing two precursors of thiamine	Schopfer [1943]
Rhodotorula sp.	thiazole	thiamine		
Marine bacterium sp.1	riboflavin	thiamine	Mutualism with precursors	Burkholder [1963]
Marine bacterium sp.2	pantothenate	thiamine		
Bacillus polymyxa	biotin	nicotinic acid	Mutualism	Yeoh et al. [1968]
Proteus vulgaris	nicotinic acid	biotin		
Auxotrophic algae	—	vitamins	Commensalism	Carlucci & Bowes [1970]
Heterotrophic bacteria	vitamins	—		
Saccharomyces cerevisiae	nicotinic acid	—	Commensalism	Shindela et al. [1965]
Proteus vulgaris	—	nicotinic acid		

TABLE 2 (cont'd)

Organism	Compound product	Compound requirement	Type of interaction	Reference
Acetobacter suboxydans	fructose (from manitos)	—	Commensalism	Chao & Reilly [1972]
Saccharomyces carlbergansis	—	fructose		
Lactobacillus plantarum	lactate (from glucose)	—	Commensalism	Lee *et al.* [1976]
Propionibacterium shermanii	—	lactate (in preference to glucose)		
Rhodopseudomonas capsulatus	carbohydrates	fatty acids	Mutualism	Okuda & Kobayoshi [1963]
heterotrophic bacteria	fatty acids	carbohydrates		
Diphtheroid	acetyl-phosphate	—	Commensalism	Nevin *et al.* [1960]
Borrelia vincenti	—	acetyl-phosphate		

pure culture of the methane-oxidizing organism. Whether or not the community exhibits a greater stability to other stresses not involving intermediates or products of methane oxidation has not been examined but the presence of more than one organism, including the peripheral species of *Flavobacterium* and *Acinetobacter* in this case, increases the possibility of coping with various stresses.

More recently Jones and Hood [1980] have demonstrated the importance these relationships may have with respect to particular metabolic activities of the primary organisms. As these authors point out "it is well recognized but not especially well documented" (and it might be added, usually ignored) that it is difficult to isolate nitrifying bacteria without accompanying heterotrophic populations and that the presence of the heterotrophs ensures better growth and increased rates of nitrification. This interaction may well be common and important for all chemolithotrophs, especially those which have an obligate mode of nutrition [Whittenbury and Kelly, 1977]. Jones and Hood [1980] described a three-membered community isolated from an estuarine environment containing an ammonium-oxidizing *Nitrosomonas* sp, associated with two heterotrophs, *Nocardia atlantica* and a species of *Pseudomonas*. The heterotrophic organisms were shown to be incapable of heterotrophic nitrification. After growth as a mixed culture the level of nitrite produced by the *Nitrosomonas* sp. was 150% greater than the organism growing in the absence of the heterotrophs. Furthermore the growth of the heterotrophs was stimulated by a factor of 10 by the presence of the chemolithotroph. It is possible that the stimulation of nitrification was due to the production of required metabolites for the *Nitrosomonas* sp. but it is more likely, however, that the stimulation was as a result of the removal of organic compounds, toxic to the *Nitrosomonas* sp. and a known problem for the growth of fastidious chemolithotrophs [Pan and Umbreit, 1972; Whittenbury and Kelly, 1977].

2.3. Class 3 Microbial Communities

Class 3 microbial communities may, as has previously been suggested (page 4), be the consequences of a number of nutritional interactions affecting the overall growth kinetics. However, it is still possible that other types of interactions, especially those involving physico-chemical properties, may improve the growth kinetic parameters of the community compared with component populations. For example, there is some evidence that floc formations influence kinetics.

2.4. Class 4 Microbial Communities

Class 4 microbial communities are frequently encountered and are characterized by a structure which is summarized in Table 3. Under the conditions in which these associations are isolated, the interactions normally establish a mutualistic relationship: that is, the survival and growth of the community depends on the concerted activity of two or more populations. However, under other growth conditions, the metabolic relationships may not be required or expressed and, for this reason, these mixtures are sometimes said to exhibit synergism or protocooperation, implying that the mutualism is not obligatory.

So far most combined metabolism communities studied have exhibited mutualism in terms of biodegradative functions (Table 4) although this is not always the case. For example, Gale [1940] demonstrated that the synthesis of putrescine from arginine was due to the concerted metabolism of *Streptococcus faecalis* and *Escherichia Coli* (Fig. 2).

Pickaver [1976] also demonstrated that combined metabolism by several organisms may result in the synthesis of a product which can accumulate. A perfusion enrichment culture, with nitriloacetate (NTA) as the

TABLE 3

The Principle of Class 4 Microbial Communities
[after Slater and Godwin, 1981]

Pathway

$$A \xrightarrow{a} B \xrightarrow{b} C \xrightarrow{c} D \xrightarrow{d} E$$

Organism	Capability
X	Requires E (or a derivative of E) for growth. No growth on A, B, C or D. Produces enzymes a and b. Intermediate C accumulates from either A or B. C and D not metabolized.
Y	Requires E (or a derivative of E) for growth. No growth on or transformation of A or B. Produces enzymes c and d. Growth on C, D and E.
X + Y	Requirement for E may be satisfied by combined metabolism of A. Growth on A, B, C, D or E. Complete pathway present since the community contains enzymes a, b, c and d.
Z	Growth on A, B, C, D or E. Complete pathway present since organism contains enzymes a, b, c and d as a result of the transfer of genes for enzymes a and b from organism. X to Y
Z^1	Growth on A, B, C, D or E. Complete pathway present since organism contains enzymes a, b, c and d as a result of transfer of genes for enzymes c and d from organism Y to X.

carbon source and sodium nitrate as the nitrogen source, produced a five-membered community consisting of three pseudomonads (*Pseudomonas* sp. strains A, B or C), a *Bacillus* sp. and a yeast. After a period of growth some of the NTA was converted to N-nitrosoimino-diacetate (NIDA), probably as a result of an interaction between intermediates of NTA metabolism, namely iminodiacetate (IDA), and nitrate metabolism, namely nitrite. Analysis of the community showed that none of the isolates alone produced NIDA when incubated with NTA and nitrate. But two two-membered mixed cultures (*Pseudomonas* strain A with *Pseudomonas* strain B or *Pseudomonas* strain A with *Bacillus* sp.) could produce NIDA from NTA and nitrate.

TABLE 4

Examples of Class 4 Microbial Communities Showing Concerted Metabolic Capabilities

Compound	Organisms	Mechanism/Comment	Reference
Diazinon (Fig. 3)	*Arthrobacter* sp. *Streptomyces* sp.	(1) Some degradation alone (2) No growth by organisms alone (3) Concerted attack for ring cleavage	Gunner & Zuckerman [1968]
Linear alkyl benzene sulphonates (LAS)	Multimembered community dominated by *Pseudomonas* spp. and *Alcaligenes* spp.	(1) No growth by pure isolates alone (2) Synergism in alkyl chain degradation (?)	Johanides & Hršak [1976]
LAS	*Pseudomonas putida* *Pseudomonas alcaligenes* *Arthrobacter globiformis* *Seratia marcescens*	(1) Highest rate of degradation by four-membered community (2) Floc structure of four-membered community important	Phillips and Hollis cited in Slater & Lovatt [1981]
Alkylphenol ethoxylates (Fig. 4)	*Nocardia* sp. *Cylindrocarpon* sp.	(1) *Nocardia* sp. for ethoxylate renewal (2) *Cylindrocarpon* sp. for ring cleavage	Baggi *et al.* [1978]
Styrene	Undefined	Unknown	Sielicki *et al.* [1978]
3,4-dichloro propionanilide (propanil)	*Penicillium piscarium* *Geotrichum candidum*	(1) *Penicillium* sp. initially metabolizes to produce 3,4-dichloroaniline (34DCA) (2) *Geotrichum* sp. condenses 34DCA to 3,3,4,4-Tetrachloroazobenzene	Bordeleau & Bartha [1968]
3-methyl heptane	*Pseudomonas* sp. *Nocardia* sp.	(1) No growth alone (2) Mechanism unknown	Wodzinski & Johnson [1968]
n-alkanes and other hydrocarbons	*Candida intermedia* *Candida lipolytica*	Unknown	Miller & Johnson [1966]
n-hexadecane	Seven-membered community including *Aeromomas* spp., *Pseudomonas* spp. + *Vibrio* spp.	Unknown	Schwartz *et al.* [1975]
Polychlorinated biphenyls	Several soil or sediment communities dominated by *Alcaligenes odorans*, *A. denitrificans* and unidentified bacterium	Unknown	Clark *et al.* [1979]
	An unidentified marine community	Unknown	Carey & Harvey [1978]
Crude oil	Undefined mixed cultures	Unknown	Horowitz *et al.* [1975]
2-(2-methyl-4-chloro)-phenoxy propionic acid	Undefined but comprising 10-12 bacteria with fluorescent pseudomonads	Unknown	Kilpi [1980]

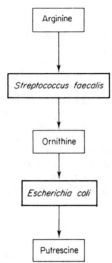

Fig.2. Conversion of arginine to putrescine by a mixture of *Streptococcus faecalis* and *Escherichia coli*.

It appeared that the *Pseudomonas* strain A was primarily responsible for NIDA synthesis but required a co-factor, which was not identified, produced by either *Pseudomonas* strain C or the *Bacillus* sp. This is similar to the interactions between two pseudomonads, described by Bates and Liu [1963], needed to produce complete lecithinase activity.

However, most of the communities described so far are a straightforward application of the principle shown in Table 3. For example, the insecticide Diazinon (O,O-diethyl-O-2-isopropyl-4-methyl-6-pyrimydyl thiophosphate) required the combined activity of an *Arthrobacter* sp. and a *Streptomyces* sp. to complete the breakdown, especially ring cleavage. Baggi et al. [1978] showed in more detail the sequential nature of the metabolic interactions to effect complete mineralization of a non-ionic surfactant (Fig. 3).

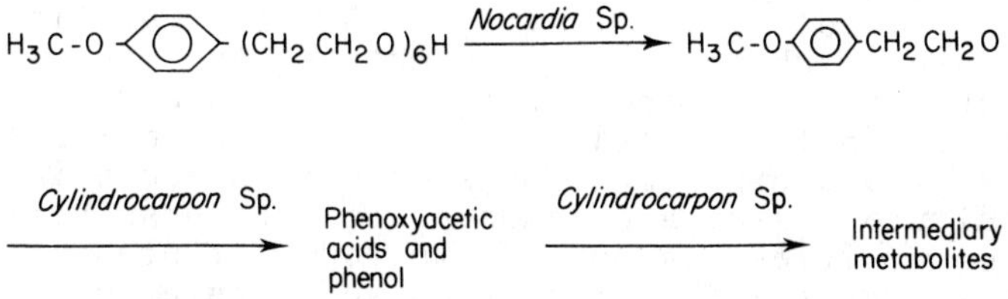

$H_3C-O-\langle O \rangle-(CH_2CH_2O)_6H \xrightarrow{\textit{Nocardia} \text{ Sp.}} H_3C-O-\langle O \rangle-CH_2CH_2O$

$\xrightarrow{\textit{Cylindrocarpon} \text{ Sp.}}$ Phenoxyacetic acids and phenol $\xrightarrow{\textit{Cylindrocarpon} \text{ Sp.}}$ Intermediary metabolites

Fig. 3. The interaction between species of *Nocardia* and *Cylindrocarpon* of a non-ionic surfactant.

2.5. Class 5 Microbial Communities

This category of microbial community constitutes a special case of the class 4 community. The specific feature of these communities is that a primary population growing on a readily metabolized compound simultaneously cometabolizes one or more compounds to yield a product which in turn supports the growth of a second population. The key feature of the system is that the second population cannot grow on the parent compound and is, therefore, completely dependant on the cometabolizing primary population. Since it is more complex to analyse reactions which do not support growth, cometabolism has been poorly studied, but it is becoming clear that these reactions may well be important in nature and the basis of many communities. A number of cometabolic communities are given in Table 5.

TABLE 5

Examples of Class 5 Microbial Communities Showing the Importance of Cometabolism

Compounds			Organisms		References
Primary substrate	Cometabolite	Cometabolite product	Primary	Secondary	
Propane	cycloalkanes	cycloketones	*Micobacterium vaccae*	Unidentified strain CY6	Beam & Perry [1973; 1974]
Cell lysis products or p-nitrophenyl breakdown products	Parathion	p-nitrophenol	*Pseudomonas stutzeri*	*Pseudomonas aeruginosa*	Daughton & Hsieh [1977]

Some attention has been given to the construction of some microbial communities involving a cometabolic stage [A.H. Filipiuk and J.H. Slater, quoted in Slater and Lovatt (1981)]. These are laboratory-based systems and it is not known if they operate in natural habitats. *Pseudomonas putida* strain PP3 contains two dehalogenases capable of dechlorinating monochloroacetic acid (MCA) yielding glycollate. *P. putida* PP3 cannot utilize glycollate which therefore accumulates and represents an ideal substrate for the growth of a secondary glycollate-utilizing population (Fig. 4). Table 6 summarizes the results of one experiment carried out in a continuous-flow culture system and demonstrates the possibility that these communities can be constructed. At present, it is uncertain why the *Flavobacterium* sp. population was at a low level despite the fact that there was adequate glycollate present which should have been utilized.

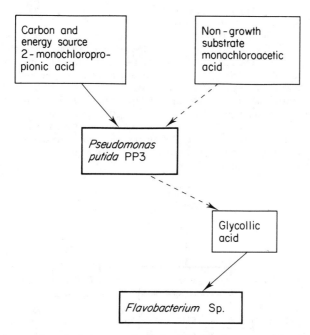

Fig. 4. A constructed two-membered microbial community involving an initial cometabolic step.

TABLE 6

The Construction of a Laboratory Two-Membered Microbial Community with a Cometabolic Stage

Carbon sources	Population Levels (organisms ml^{-1})		Cl$^-$ concentration (µmol ml^{-1})	Glycollate concentration (µmol ml^{-1})	Comment
	Pseudomonas putida	*Flavobacterium* sp.			
2MCPA alone	0.5 to 3.0 x 10^8	0	6 — 10	0	Growth of primary population only.
2MCPA and MCA	0.5 to 3.0 x 10^8	0	16 — 18	10	Dechlorination of two substrates by primary organism.
2MCPA and MCA	0.5 to 3.0 x 10^8	3.0 to 6.0 x 10^5	16 — 18	9	Growth of secondary organism on excreted glycollate.

2.6. Class 6 Microbial Communities

Under anaerobic conditions fermentative organisms require a sink to dispose of excess reducing power which is normally accomplished by the reduction of one or more of the terminal products of a fermentative pathway. A number of tight associations have been described in which a second organism acts as the election sink, thus there is a process of interspecies hydrogen ion transfer. These have been extensively discussed elsewhere [Wolin, 1975, 1976, 1977] with the "organism" *Methanobacillus omelianskii* representing one of the better known communities (Fig. 5).

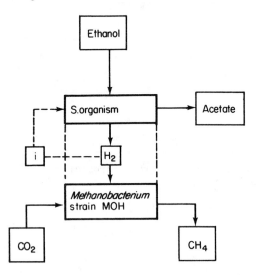

Fig. 5. *Methanobacillus omeliamskii* (after Slater [1978]).

2.7. Class 7 Microbial Communities

Finally, many continuous-flow culture enrichments result in the isolation of highly stable microbial communities which contain more than one species capable of growing on the sole carbon and energy source which is provided: that is, there is more than one primary utilizer. In addition there are often a number of additional populations — secondary organisms — which are unable to metabolize the primary substrate but which are nevertheless stable members of the community. We have isolated a number of such communities growing on the herbicides Dalapon [Senior et al., 1976] (Fig. 6), Lontrel [Lovatt et al., 1978], Fenuron, Monuron and Dicamba [Parkes et al.,1980; Minney et al., 1980] and others have isolated similar communities growing on compounds such as benzoic acid [Cossar et al., 1981]. In

one respect these microbial communities are a little
surprising since it would be expected that in an open
culture system after a suitable enrichment period a pure
culture would result. On the other hand, as has been
argued in this chapter, interactions of a variety of
types frequently occur and it is therefore to be expected
that such assemblages should be obtained. One of the
present difficulties with this type of community is that
nature of the interactions, especially those involving
the primary utilizers, have not been elucidated: they
must occur in order to stabilize free competition for a
single growth substrate, especially if it is present at
a growth-limiting concentration.

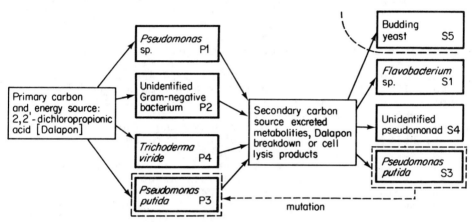

Fig. 6. The Dalapon microbial community (after Senior et al. [1976]).

3. Features and Problems Associated with Microbial Communities

In this section I wish to catalogue a number of
features of microbial communities which may be significant and to discuss some of the difficulties currently
encountered with the studies of microbial communities.

3.1. *Do They Exist in Natural Habitats?*

The alternative question is to ask whether or not
microbial communities constitute artifacts of a particular enrichment procedure with no significance to the
overall microbial activity in a particular habitat.
Furthermore, if they do occur, bearing some relationship
to the form observed in laboratory systems, what other
environmental or biological factors affect their
characteristics and how? The fact that a particular
stable, interacting assemblage of microbes can be
isolated from a particular sample shows, at the least,
that the potential for forming such a community exists.

In some cases it is a simple matter to demonstrate the physical relationships between different microbial species in natural samples and make direct comparisons with the mixed cultures isolated in the laboratory. For example, some of the tight microbial communities, particularly those involving photosynthetic bacteria and sulphate-reducing bacteria [Gray et al., 1973]. In other cases the obligatory nature of the relationship has led to the erroneous conclusion that a pure culture has been obtained when in fact the culture was a tightly coupled association: for example, the organism known as "*Chloropseudomonas ethylica*" subsequently was discovered to be a mixture of a *Chlorobium* sp. and a *Desulfovibrio* sp. [Gray et al., 1973]. It is also obvious that many loose associations can be directly observed: for example, epiphytic bacteria and diatoms growing attached to the surface of unicellular and filamentous eukaryotic algae.

There is little evidence to suggest that particular microbial communities do exist and function in nature. For example, the Diazinon community [Gunner and Zuckerman, 1968] (Fig. 3; Table 4) was observed as the result of a selective enrichment and increase in numbers of *Arthrobacter* sp. and *Streptomyces* sp. in soil treated with Diazinon. This is not to say, and indeed it seems unlikely, that prior to the Diazinon treatment the *Arthrobacter* and the *Streptomyces* species grew and functioned as a two-membered community. It is more likely that particular populations rapidly associate to form a particular community to meet a particular metabolic challenge, especially with respect to the degradation of xenobiotic compounds. If this is so then a given species could have one function in one community and a different function in another. To date there have been no studies undertaken to examine this possibility and, indeed, it will be technically extremely difficult to achieve with any degree of confidence. It will also be a difficult task to determine the function and activity of a given community in its natural habitat.

3.2. *Why Do They Occur?*

There are likely to be several answers to this question depending on the type of microbial community under consideration. At one level microbial communities are the inevitable consequence of the flow of energy, carbon and other elements between different organisms at the same time or different trophic levels. This is particularly true of sequential, commensal relationships in which the end products of the metabolism of one organism represent an energy or material resource to be exploited by other organisms with different capabilities. Thus an organism which has to produce an organic compound as the end product of its fermentation will be excreting a compound utilizable by another organism. Similarly organisms which maximize their own energetics processes through the action of slip reactions [Neijssel and Tempest, 1976] are also generating materials which may be exploited by others. These mechanisms, of course, ensure the most efficient (for a given set of conditions) utilization of all available resources.

Another answer may be that microbial communities exist and are rapidly selected for in order to exploit a particular resource. This seems clearly to be the case with respect to complex compounds and xenobiotic compounds where the complete metabolic potential to metabolize a given compound simply does not exist in the genetic complement of a single species (see pages 8-11).

Microbial communities could also exist in order to maximize factors such as the rate of growth or the yield of all the organisms involved in the community. This is clearly the case with those communities structured on the basis of the removal of toxic metabolic products. In another example, the rate of Dalapon breakdown by the microbial community is greater than the sum of the individual rates of the primary organisms growing alone [Senior, 1977].

However, one problem with this argument is that organisms sensitive to particular products could have evolved a resistance to the inhibition instead of evolving a relationship with a second organism which might compete with the primary organism for other resources. For example, in the four-membered methanol-inhibition community [Wilkinson et al., 1974], the *Hyphomicrobium* species, even though it is present at low levels, still has a requirement for nitrogen, phosphorus, etc. Under conditions in which methane is the limiting substrate (i.e., the conditions of the original enrichment), these common requirements are not likely to be a significant problem. Nevertheless, the presence of the secondary population constitutes a drain on resources which otherwise might be utilized by the primary production yielding a larger number of organisms or a population capable of faster growth. One answer to this situation may be that these relationships have evolved, instead of more efficient, resistant, primary populations growing alone, to provide other characteristics which are more valuable, such as stability and the capacity to deal with transient stress conditions (such as the sudden appearance of a high concentration of methanol) more effectively than by a pure culture.

At the present, without any detailed experimental evidence to support or disprove the idea of better stability and capacity to respond to transient conditions for microbial communities compared with pure cultures, it is not very clear why communities of the class 1 type seem to be so common. Simply at the level of nutritional needs it does not seem to be an efficient strategy to adopt to be dependent upon growth requirements synthesized and excreted by another organism. Furthermore it can be readily shown that organisms which overproduce and excrete a metabolite grow at slower rates and have poorer yields [Baich and Johnson, 1968]. Thus there is a double disadvantage to the donor and the recipient. It has been shown, however, that auxotrophic strains which have their growth requirement met exogenously do have a competitive advantage, in terms of rate of growth, over the isogenic, prototrophic strain [Zamenhoff and Eichhorn, 1967; Mason and Slater, 1979]. It appears, therefore, that under certain conditions — i.e., high external concentrations of the growth requirement — there is some advantage to not carrying the genes for, or expressing the pathway for, the synthesis of a particular metabolite. Whether this advantage still exists when the required metabolite is present at low external concentrations is doubtful since prototrophic revertants of a tyrosine auxotrophic strain of *Escherichia coli* rapidly outcompete the auxotroph when they were grown under conditions of tyrosine limitation [Mason and Slater, 1979]. In many cases these communities seem to revolve around only one or two requirements. For example, in the *Saccharomyces cerevisiae/lactobacillus casei* mixture [Megee et al., 1972] the predominant, if not the sole, requirement is for riboflavin. Compared with the genetic composition the lactobacillus already possesses, the addition of the riboflavin biosynthetic capability seems trivial, even if the extra genetic material does marginally reduce the maximum rate of growth. This appears even more curious when it is realized that a stable association exists under glucose limited conditions. When riboflavin is added to the growth medium, the lactobacillus rapidly outcompetes the yeast. That is, the rate of supply of riboflavin by the yeast restricts the lactobacillus' growth rate. This analysis seems to indicate that there is an advantage to the association, so far unspecified, which is more significant than a simple improvement in the organisms growth rate and is worthy of further detailed study.

3.3. *Characterization of Isolated Microbial Communities*

A technical problem which needs to be stated and borne in mind in mixed culture studies, is the question of identifying the component members of the mixture. Analysis requires the disruption of the community and attempts to grow the component populations alone. Whether all the component populations are always recognized is doubtful, leaving the concern that an important member of the community has not been isolated. Associated with this is the difficulty of recognizing the presence of and role of a component population which may be present at levels which are orders of magnitude lower than the dominant members of the community.

3.4. Adaptation in Microbial Communities and Through the Formation of Microbial Communities

Another property of microbial communities likely to have a wide significance is their ability to adapt, at the genetic level, more readily to novel environmental changes or challenges. This can occur in two ways.

Firstly, adaptations can occur within an established microbial community resulting in a change in the structure of the community or the appearance of an organism with a new metabolic capability. For example, Senior et al. [1976] showed that within a stable association, utilizing the carbon source originally used to select the community, a secondary organism (unable to metabolize the sole carbon source initially) evolved into a primary organism after prolonged culture (Fig. 7). This was only possible in this particular system so long as the secondary organism (*Pseudomonas putida* S3) was retained within the structure of the community for a sufficient period of time to adapt. In other words the community provided a permissive environment by providing growth substrates to sustain the growth of the secondary organism [Senior et al., 1976]. This can be minimized using a pure culture of *P.putida* S3 only by providing a second, readily metabolized substrate to sustain the population which may then be continuously exposed to the novel carbon source. Another interesting point about this particular system is that the selection of a Dalapon-utilizing strain of *Pseudomonas putida* S3 occurred despite the presence of other Dalapon-utilizing organisms. The adaptation seems to be due to the expression of dehalogenase enzymes, probably as a result of a regulatory mutant [Slater et al., 1979; A.J. Weightman, A.L. Weightman and J.H. Slater, unpub. obs.].

Similarly it has been shown that a microbial community isolated using picolinic acid (pyridine-2-carboxylic acid) as the carbon source could be adapted to metabolize partially a chlorinated analogue, 3,6-dichloropicolinic acid. The community contained three primary organisms: *Pseudomonas aeruginosa*, *Alcaligenes faecalis* and a second *Alcaligenes* sp., and three secondary organisms: *Bacillus licheniformis*, *Rhodococcus* sp. and *Corynebacterium* sp. [Lovatt et al., 1978; Slater and Lovatt, 1981]. After an appropriate sequence of adaptation events a new community was selected with a major increase in the proportion of the *Alcaligenes* sp. in the complete community and the capacity to grow in the presence of 3,6-dichloropicolinic acid, possibly by a novel capability to hydroxylate the chlorinated compound.

Secondly, adaptations can occur through the establishment of an interacting microbial community in the first place. For example, it might be possible to observe the selection of organism Z or organism Z^1 as the result of the transfer of genetic information between members of

the initial two-membered community X and Y (Table 3).
For this to occur appropriate gene transfer mechanisms
compatible with the organisms involved must operate and
the evolved organism ought, preferably, to be at a competitive advantage over the primary community. Such
events probably occur readily in natural habitats but so
far there has been little experimental evidence with
laboratory systems to demonstrate such a sequence of
events. Knackmuss and co-workers [Hartman et al.,1979;
Reineke and Knackmuss, 1979] have provided some evidence
with plasmid-mediated transfer of genes involved in the
catabolism of halogenated aromatic compounds to suggest
that the evolution of new organisms may stem from an
initial mixed culture of organisms.

4. Conclusion

This chapter has demonstrated the potential for many
different types of interacting microbial systems. What
is lacking at the present time is a systematic analysis
of factors such as their frequences and significance and
their general importance in nature. Furthermore there is
a considerable gap with respect to the exploitation of
stable microbial communities for biosynthetic processes
and the formation of economically useful products.

References

Baggi, G., Beretta, L., Galli, E., Scolastico, C. and Treccani, V. (1978). Biodegradation of alkylphenol polyethoxylates. In "The Oil Industry and Microbial Ecosystems" (eds. K.W.F. Chater and H.J. Somerville), pp.129-136. London: Heyden and Sons.
Baich, A. and Johnson, M. (1968). Evolutionary advantage of control of a biosynthetic pathway. *Nature, London* **218**, 464-465.
Bates, J.L. and Liu, P.V. (1963). Complementation of lecithinase activities in closely related pseudomonads: its taxonomic implications. *Journal of Bacteriology* **86**, 585-592.
Beam, H.W. and Perry, J.J. (1973). Co-metabolism as a factor in microbial degradation of cycloparaffinic hydrocarbons. *Archiv für Mikrobiologie* **91**, 87-90.
Beam, H.W. and Perry, J.J. (1974). Microbial degradation of cycloparaffinic hydrocarbons via cometabolism and commensalism. *Journal of General Microbiology* **82**, 163-169.
Bordeleau, L.M. and Bartha, R. (1968). Ecology of a pesticide transformation: synergism of two soil fungi. *Soil Biology and Biochemistry* **3**, 281-284.
Brock, T.D. (1969). Microbial growth under extreme conditions. In "Microbial Growth" (eds. P.M. Meadow and S.J. Pirt), pp.15-41. Cambridge: Cambridge University Press.
Bull, A.T. and Brown, C.M. (1979). Continuous culture applications to microbial biochemistry. In "Microbial Biochemistry" (ed. J.R. Quayle), pp. 177-226. Baltimore: University Park Press.
Bull, A.T. (1980). Biodegradation: some attitudes and strategies of microorganisms and microbiologists. In "Contemporary Microbial

Ecology" (eds. D.C. Ellwood, J.N. Hedger, M.J. Latham, J.M. Lynch and J.H. Slater), pp.107-136. London: Academic Press.
Bungay, H.R. and Bungay, M.L. (1968). Microbial interactions in continuous culture. *Advances in Applied Microbiology* **10**, 269-290.
Burkholder, P.R. (1963). Some nutritional relationships among microbes of sea sediments and waters. In "Symposium on Marine Microbiology", (ed. C.H. Oppenheimer), pp. 133-150. Springfield, Illinois: C.C. Thomas.
Carey, A.E. and Harvey, G.R. (1978). Metabolism of polychlorinated biphenyl by marine bacteria. *Bulletin of Environmental Contamination and Toxicology* **20**, 527-534.
Carlucci, A.F. and Bowes, P.M. (1970). Vitamin production and utilization by phytoplankton in mixed culture. *Journal of Phycology* **6**, 393-400.
Chao, C-C. and Reilly, P.J. (1972). Symbiotic growth of *Acetobacter suboxydans* and *Saccharomyces carlbergansis* in a chemostat. *Biotechnology and Bioengineering* **14**, 75-92.
Clark, R.R., Chian, E.S.K. and Griffin, R.A. (1979). Degradation of polychlorinated biphenyls by mixed microbial cultures. *Applied and Environmental Microbiology* **37**, 680-685.
Cossar, D., Brown, C.M. and Watkinson, R.J. (1981). Some properties of a marine microbial community utilizing bezoate. *Society for General Microbiology Quarterly*, **8**, p.147.
Cremieux, A., Chevalier, J., Combert, M., Dumeuil, G., Parlouar, D. and Ballerini, D. (1977). Mixed culture of bacteria utilizing methanol for growth. I. Isolation and identification. *European Journal of Applied Microbiology* **4**, 1-9.
Daughton, C.G. and Hsieh, D.P. (1977). Parathion utilization by bacterial symbionts in a chemostat. *Applied and Environmental Microbiology* **34**, 175-184.
Gale, E.F. (1940). The production of amines by bacteria. III. The production of putrescine from arginine by *Bacterium coli* in symbiosis with *Streptococcus faecalis*. *Journal of Biochemistry* **34**, 853-857.
Gray, B.H., Fowler, C.F., Nugent, N.A., Rigopoulos, N. and Fuller, R.C. (1973). Reevaluation of *Chloropseudomonas ethylica*. *International Journal of Systematic Bacteriology* **23**, 256-264.
Gunner, H.B. and Zuckerman, B.M. (1968). Degradation of 'Diazinon' by synergystic microbial action. *Nature, London* **217**, 1183-1184.
Hartmann, J., Reineke, W. and Knackmuss, H-J. (1979). Metabolism of 3-chloro-, 4-chloro- and 3,5-dichlorobenzoate by a pseudomonad. *Applied and Environmental Microbiology* **37**, 421-428.
Horowitz, A., Gutnick, D. and Rosenberg, E. (1975). Sequential growth of bacteria on crude oil. *Applied Microbiology* **30**, 10-19.
Jensen, H.L. (1957). Decomposition of chloro-substituted aliphatic acids by soil bacteria. *Canadian Journal of Microbiology* **3**, 151-164.
Johanides, V. and Hršak, D. (1976). Changes in mixed bacterial culture during linear alkylbenzene-sulfonate (LAS) biodegradation. In "Abstracts of Papers, Fifth International Fermentation Symposium (ed. H. Dellweg), p.124. Berlin: Verlag Versuchs-und Lehranstalt für Spiritusfabrikation und Fermentationstechnologie.
Jones, R.D. and Hood, M.A. (1980). Interaction between an ammonium-oxidizer, *Nitrosomonas* sp., and two heterotrophic bacteria.

Microbial Ecology **6**, 271-275.

Kilpi, S. (1980). Degradation of some phenoxy acid herbicides by mixed cultures of bacteria isolated from soil treated with 2-(2-methyl-4-chloro)phenoxypropionic acid. *Microbial Ecology* **6**, 261-270.

Koch, R. (1881). Zur Untersuchung von pathogenen organismen. *Mittheilungen aus dem Kaiserlichen Gesundheitsamte* **1**, 1-48.

Kuenen, J.G. and Harder, W. (1981). Microbial competition in continuous culture. In "Experimental Microbial Ecology" (eds. R.G. Burns and J.H. Slater). Oxford: Blackwell Scientific Publications.

Lee, I.H., Fredrickson, A.G. and Tsuchiya, H.M. (1976). Dynamics of mixed cultures of *Lactobacillus plantarum* and *Propionibacterium shermanii*. *Biotechnology and Bioengineering* **18**, 513-526.

Lovatt, D., Slater, J.H. and Bull, A.T. (1978). The growth of a stable mixed culture on picolinic acid in continuous-flow culture. *The Society for General Microbiology Quarterly* **6**, 27-28.

Mason, T.G. and Slater, J.H. (1979). Competition between an *Escherichia coli* tyrosine auxotroph and prototrophic revertant in glucose- and tyrosine-limited chemostats. *Antonie van Leewenhoek* **45**, 253-263.

Meers, J.L. (1973). Growth of bacteria in mixed cultures. *CRC Critical Reviews in Microbiology* **2**, 139-184.

Megee, P.D., Drake, J.F., Fredrickson, A.G. and Tsuchiya, H.M. (1972). Studies in intermicrobial symbiosis. *Saccharomyces cerevisiae* and *Lactobacillus casei*. *Canadian Journal of Microbiology* **18**, 1733-1742.

Miller, T.L. and Johnson, M.J. (1966). Utilization of gas oil by a yeast culture. *Biotechnology and Bioengineering* **8**, 567-580.

Minney, S., Parkes, R.J., Slater, J.H. and Bull, A.T. (1980). The degradation of Fenuron by estuarine microbial communities in laboratory microcosms. "Abstracts of the Second International Symposium on Microbial Ecology", pp. 117-118.

Neijssel, O.M. and Tempest, D.W. (1976). The role of energy-spilling reactions in the growth of *Klebsiella aerogenes* NCTC 418 in aerobic chemostat culture. *Archives for Microbiology* **110**, 305-311.

Nevin, A., Hampp, E.G. and Duey, B.V. (1960). Interaction between *Borrelia cincentii* and an oral diphtheroid. *Journal of Bacteriology* **80**, 783-786.

Nurmikko, V. (1954). Symbiosis experiments concerning the production and biosynthesis of certain amino acids and vitamins in association of lactic acid bacteria. *Annals Academiae Scientianim Fenniniae* **54**, 7-58.

Nurmikko, V. (1956). Biochemical factors affecting symbiosis among bacteria. *Experimentia* **12**, 245-249.

Okuda, A. and Kebayoshi, M. (1963). Symbiotic relationship between *Rhodopseudomonas capsulatus* and *Azotobacteri nivelaudii*. *Mikorobiologya* **32**, 792-804.

Osman, A., Bull, A.T. and Slater, J.H. (1976). Growth of mixed microbial populations on oranol in continuous culture. In "Abstracts of Papers, Fifth International Fermentation Symposium" (ed. H. Dellweg), p. 124. Berlin: Verlag Versuchs-und Lehranstalt für Spiritusfabrikation und Fermentations-technologie.

Pan, P. and Umbreit, W.W. (1972). Growth of obligate autotropic bacteria on glucose in a continuous flow-through apparatus. *Journal of Bacteriology* **109**, 1149-1155.

Parkes, R.J., Minney, S.F., Slater, J.H. and Bull, A.T. (1980). Effects of enrichment conditions on the isolation of microbial communities from the estuarine environment. "Abstracts of the Second International Symposium on Microbial Ecology", p.117.

Pfennig, N. (1978). Syntrophic associations and consortia with phototrophic bacteria. In "Abstracts of the XII International Microbiology Congress of Microbiology", p.16.

Pickaver, A.H. (1976). The production of n-nitrosoiminodiacetate from nitrilotriacetate and nitrate by microorganisms growing in mixed culture. *Soil Biology and Biochemistry* **8**, 13-17.

Reineke, W. and Knackmus, H.-J. (1979). Construction of haloaromatic utilizing bacteria. *Nature, London* **277**, 385-386.

Schopfer, W.H. (1943). "Plants and Vitamins". Waltham: Chronica Botanica.

Schwarz, J.R., Walker, J.D. and Colwell, R.R. (1975). Deep-sea bacteria: growth and utilization of n-hexadecane at *in situ* temperature and pressure. *Canadian Journal of Microbiology* **21**, 682-687.

Senior, E., Bull, A.T. and Slater, J.H. (1976). Enzyme evolution in a microbial community growing on the herbicide Dalapon. *Nature, London* **263**, 476-479.

Senior, E. (1977). Characterization of a microbial association growing on the herbicide Dalapon. Ph.D. thesis, University of Kent.

Shindela, A., Bungay, H.R. and Krieg, N.R. (1965). Mixed culture interactions. I. Commensalism of *Proteus vulgaris* and *Saccharomyces cerevisiae* in continuous culture. *Journal of Bacteriology* **89**, 693-696.

Sielicki, M., Focht, D.D. and Martin, J.P. (1978). Microbial transformations of styrene and (^{14}C)-styrene in soil and enrichment cultures. *Applied and Environmental Microbiology* **35**, 124.

Slater, J.H. and Bull, A.T. (1978). Interactions between microbial populations. In "Companion to Microbiology" (eds. A.T. Bull and P.M. Meadow), pp.181-206. London: Longman.

Slater, J.H. (1978). The role of microbial communities in the natural environment. In "The Oil Industrial and Microbial Ecosystems" (eds. K.W.A. Chater and H.J. Somerville), pp. 137-154. London: Heydon & Sons.

Slater, J.H., Lovatt, D., Weightman, A.J., Senior, E. and Bull, A.T. (1979). The growth of *Pseudomonas putida* on chlorinated aliphatic acids and its dehalogenase activity. *Journal of General Microbiology* **114**, 125-136.

Slater, J.H. and Godwin, D. (1980). Microbial adaptation and selection. In "Contemporary Microbial Ecology" (eds. D.C. Ellwood, J.N. Hedger, M.J. Latham, J.M. Lynch and J.H. Slater), pp.137-160. London: Academic Press.

Slater, J.H. and Lovatt, D. (1981). Biodegradation and the significance of microbial communities. In "Biochemistry of Microbial Degradation" (ed. D.T. Gibson). New York: Marcel Dekker.

Stirling, L.A., Watkinson, R.J. and Higgins, I.J. (1976). The microbial utilization of cyclohexane. *Proceedings of the Society for General Microbiology* **4**, 28.

Whittenbury, R. and Kelly, D.P. (1977). Autotrophy: a conceptual phoenix. In "Microbial Energetics" (eds. B.A. Haddock and W.A. Hamilton), pp.121-149). Cambridge: Cambridge University Press.

Whittenbury, R. (1978). Bacterial nutrition. In "Essays in Microbiology" (eds. J.R. Norris and M.H. Richmond), 16/1-16/32. Chichester: John Wiley & Sons.

Wilkinson, T.G., Topiwala, H.H. and Hamer, G. (1974). Interactions in a mixed bacterial population growing on methane in continuous culture. *Biotechnology and Bioengineering* **16**, 41-49.

Wodzinski, R.S. and Johnson, M.J. (1968). Yields of bacterial cells from hydrocarbons. *Applied Microbiology* **16**, 1886-1891.

Wolin, M.J. (1975). Interactions between the bacterial species of the rumen. In "Digestion and Metabolism in the Ruminant" (eds. I.W. McDonald and A.C.I. Warner), pp.134-148. Armidale, Australia: University of New England Publishing.

Wolin, M.J. (1976). Interactions between hydrogen-producing and methane-producing species. In "Symposium on Microbial Production and Utilization of Gases" (eds. H.G. Schlegel, G. Gottschalk and N. Pfennig), pp.141-150. Gottingen: E. Goltze KG.

Wolin, M.J. (1977). The rumen fermentation: a model for microbial interactions in anaerobic ecosystems. *Advances in Microbial Ecology* **3**, 44-77.

Zamenhof, S. and Eichhorn, H.H. (1967). Study of microbial evolution through loss of biosynthetic functions: establishment of "defective" mutants. *Nature, London* **216**, 456-458.

Chapter 2

MIXED CULTURE KINETICS

MICHAEL J. BAZIN

Microbiology Department, Queen Elizabeth College, Campden Hill Road, London, UK

1. Introduction

A microbial culture in which more than one species is growing is complicated because the state variables of the system are changing not only with respect to at least one independent variable, usually time, but with respect to each other as well. In order to study such a system it is often helpful to construct a conceptual framework in the form of a mathematical model. Usually these models are based on simplifying assumptions or can be applied only to idealized or very specific situations. The primary aim of such an approach is to understand how the system works. A prediction of expected behaviour is a secondary consideration in a scientific sense but, of course, of considerable interest to an engineer who might wish to optimize an industrial process with which the system is associated. Whether models of microbial interactions are constructed for their heuristic or predictive values, central to their efficacy are the kinetic functions which describe the way in which the species involved interact with each other. In this chapter I have attempted to define microbial interactions in terms of some of the mathematical expressions which have been suggested for these functions and describe some of the experimental and analytical methods which have been used to investigate mixed culture kinetics

2. Definitions

There seems to be considerable disagreement between ecologists about the names which should be used to describe associations between different species of organisms. For example, some workers regard symbiosis as meaning any type of association but others regard it as only correctly applied to cases where some mutual benefit is accrued by the two populations. In order to avoid the semantics associated with this sort of discussion I will use two equations representing a generalized interaction to aid in defining the terms I employ.

Consider a well-mixed microbial culture containing two species at population densities X_1 and X_2. In the absence of immigration or mutation, the system may be represented as follows:

$$\frac{dX_1}{dt} = (\alpha_1 + \beta_{11}X_1 + \beta_{12}X_2)X_1 \qquad (1)$$

$$\frac{dX_2}{dt} = (\alpha_2 + \beta_{22}X_2 + \beta_{21}X_1)X \qquad (2)$$

These are the so-called Lotka-Volterra equations which have been used extensively in theoretical ecology to investigate multispecies ecosystems.

Usually the coefficients, α and β are taken to be constants but unless indicated otherwise I will consider them to be functions. The α_i terms may be regarded as specific growth rate functions so for α_1 = constant, exponential growth is assumed. For:

$$\alpha_i = \frac{\alpha_{mi} S}{K_i + S} \qquad (3)$$

Monod saturation kinetics are implied where S is the concentration of growth-limiting nutrient and α_{mi} and K_i are the associated growth constants.

The second term in each of the equations, β_{ii}, represents a factor which regulates the population density in some way. For example with $\alpha_1 = r$, $\beta_{11} = -r/k$ and $\beta_{12} = 0$, Equation (1) becomes:

$$\frac{dX_1}{dt} = (r - \frac{r}{k} X_1)X_1 \qquad (4)$$

This is the Verhulst-Pearl logistic equation where, in ecological terms, r is the intrinsic rate of increase and k is the carrying capacity of the environment. Microbiologically, r is the specific growth rate and k represents the amount of biomass that is obtained after growth ceases. β_{11} in the logistic model therefore determines growth yield. The functions might be used also to represent the effect of dilution immigration or death. In order to represent a chemostat, for example, we let:

$$\beta_{ii} = D/X_i \qquad (5)$$

where D is the dilution rate.

Fredrickson [1977] classified microbial interactions as indirect if no physical contact between the species is involved but some non-living part of the environment is a necessary intermediary. For example, a product of one or both species may inhibit or promote the growth of the other. A direct interaction is one which involves direct contact between species. In terms of Equations (1) and (2) direct interactions may be thought of as those involving the β_{ij} ($i \neq j$) functions as they are associated with the product of X_1 and X_2 although, as I indicate later, representation of direct interactions is not limited to this function.

If the presence of one species promotes growth of the other, the interaction is said to be positive. In the case of a direct interaction involving the β_{ij} terms, this implies:

$$\beta_{12} > 0$$
and/or
$$\beta_{21} > 0$$

A negative direct interaction occurs when:

$$\beta_{12} < 0$$
and/or
$$\beta_{21} < 0$$

Indirect interactions are either positive or negative depending on whether the functions concerned are either increasing or decreasing with respect to the environmental factor involved. For example, one species might produce a substance which decreases the specific growth rate of the other, in which case a negative indirect interaction is said to take place.

Specific types of microbial associations may be defined in more detail by specifying the functions in Equations (1) and (2). However, as I show later in this chapter, inconsistencies arise when attempts are made to use equations of the Lotka-Volterra type to classify the interactions that can occur between microorganisms.

3. Competition

There is confusion and disagreement as to what competition between microorganisms is. I shall quote from three comparatively recent publications. Harrison [1978] considers competition to occur "when the growth of two or more species is limited by a common factor ..." Slater and Bull [1978] are more specific in their definition when they say that competition results "when both populations are limited, either in terms of growth rate or final population size, by a common dependence on an external factor required for growth". The kinetic analysis of competition carried out by these latter authors is restricted by what they call a simplifying assumption to competition for a limiting amount of an essential nutrient. Unfortunately, they then go on to discuss in theoretical terms what happens when two populations are grown together in a closed environment in which all nutrients are in excess. That is, just those conditions in which, under the assumptions they make, no competition will occur. The reason why their analysis does not reveal this inconsistency is that they represent the growth rate functions of both competing organisms as being constants. As pointed out by Megee et al. [1972] for an analysis of the situation attempted by Slater and Bull [1978] it is necessary to treat the specific growth rates of the competing organisms as functions of the limiting nutrient for which they are competing. According to the equations used by Slater and Bull [1978], reproduced by Slater [1979], no competition takes place between the microbial populations represented and, in fact, the species concerned could just as well be growing in different culture flasks.

Fredrickson [1977] regards competition as an indirect interaction which has a negative effect on both populations. If a substance required by both species is removed from the environment he defines the situation as resource-type competition while secretion of substances inhibitory to each other he calls interference-type competition. In both cases it is assumed that what is competed for is a chemical species; as I shall illustrate later, this need not necessarily be the case.

Accepting Frederickson's definition of competition as an indirect, negative interaction, in terms of Equations (1) and (2), this implies $\beta_{12} = \beta_{21} = 0$. Growth rate-limiting resource-type competition occurs when $\alpha_1 = \alpha_1(S)$ and $\alpha_2 = \alpha_2(S)$ where S is the concentration of limiting nutrient for which the two species are competing. Interference-type competition may affect the growth rate functions so that $\alpha_1 = \alpha_1(P_2)$ and $\alpha_2 = \alpha_2(P_1)$, where P_1 and P_2 are the inhibitory products of the first and second populations respectively.

The most often quoted model of microbial competition utilizes the Monod relationship to represent resource-type growth limitation. For a chemostat culture:

$$\frac{dX_1}{dt} = \frac{\alpha_{m1} S X_1}{K_1 + S} - D X_1 \qquad (6)$$

$$\frac{dX_2}{dt} = \frac{\alpha_{m2} S X_2}{K_2 + S} - D X_2 \qquad (7)$$

$$\frac{dS}{dt} = D(S_r - S) - \frac{\alpha_{m1} S X_1}{Y_1(K_1+S)} - \frac{\alpha_{m2} S X_2}{Y_2(K_2+S)} \qquad (8)$$

where Y_1 and Y_2 are the growth yields (assumed to be constant) of the two populations and S_r is the concentration of limiting nutrient in the medium entering the culture. The dynamic equations for a batch culture are, of course, a special case of these equations where $D = 0$. Analysis of Equations (6) to (8) reveals that both populations can survive only under very special circumstances. By survival I mean that both populations remain under steady state conditions at densities $\tilde{X}_1$ and $\tilde{X}_2$ and nutrient concentration $\tilde{S}$. Steady state implies:

so that
$$\frac{dX_1}{dt} = \frac{dX_2}{dt} = \frac{dS}{dt} = 0$$

$$D = \frac{\alpha_{m1} \tilde{S}}{K_1 + \tilde{S}} \quad \frac{\alpha_{m2} \tilde{S}}{K_2 + \tilde{S}} \qquad (9)$$

and

$$S_r = S + \frac{\tilde{X}_1}{Y_1} + \frac{\tilde{X}_2}{Y_2} \qquad (10)$$

From (9),

$$\tilde{S} = \frac{\alpha_{m2} K_1 - \alpha_{m1} K_2}{\alpha_{m1} - \alpha_{m2}} \qquad (11)$$

which relationship specifies a dilution rate, D', at which it is possible for the species to coexist:

$$D' = \frac{\alpha_{m2} K_1 - \alpha_{m1} K_2}{K_1 - K_2} \qquad (12)$$

However, as D' must be a positive number then either:

$$K_1 > K_2$$
and
$$\alpha_{m2} K_1 > \alpha_{m1} K_2$$

or
$$K_2 > K_1$$
and
$$\alpha_{m1} K_2 > \alpha_{m2} K_1$$

These conditions are stringent; not only do the growth constants of the two species have to conform to the above inequalities, which are illustrated graphically in Fig. 1, but in addition the dilution rate of the system must be that stipulated by Equation (12).

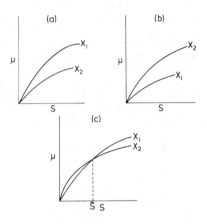

Fig. 1. Possible relationships between the specific growth rate functions of two competing species on the assumption of Monod kinetics. S is limiting substrate concentration and X_1 and X_2 the population densities of the populations. In (a) population X_1 always grows faster than population X_2 and so will outcompete the second species. In (b) the reverse is true while in (c) at limiting nutrient concentrations $\tilde{S}$ the two populations have an identical specific growth rate. In chemostat culture both species can survive if the nutrient concentration can be maintained at this value by running the system at the dilution rate defined by Equation (12).

The dynamics of competition in a chemostat culture under the assumption of Monod kinetics is illustrated diagrammatically in Fig. 2. These results were obtained by numerically solving Equations (6) to (8). Simulations of this sort give an indication of how the dependent variables in the system approach steady state and also the values of the two steady state population densities, quantities not obtainable directly from steady state analysis. In Fig. 3 the product of several simulations are summarized in a form which shows the dependence of the two population densities on their inoculation sizes and also the linear interdependence between $\tilde{X}_1$ and $\tilde{X}_2$ such that the sum of these two steady state variables is constant. This together with Equation (10) implies that $\tilde{S}$, the steady state limiting nutrient concentration, is fixed.

The discussion above applies only to the rather ephemeral condition in which a chemostat can be maintained precisely at the dilution rate given in Equation (12). Technically this condition would be very difficult

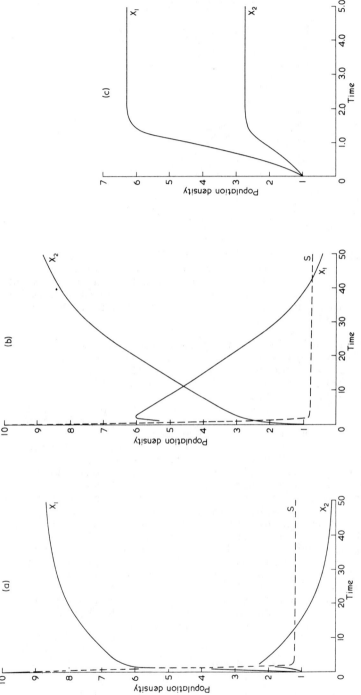

Fig. 2. Simulated results of competition in a chemostat corresponding to the relationships between specific growth rate and limiting substrate shown in Fig. 1. In all cases the initial nutrient concentration was 10 arbitrary units and the initial biomass densities for both populations were 1 arbitrary unit. In (a) population X_1 survives and population X_2 washes out, while in (b) the reverse is true. In (c) the dilution rate is maintained at the value defined in Equation (12) and both populations survive.

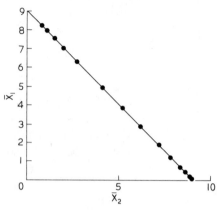

Fig. 3. Steady state biomass densities of competing species in chemostat culture calculated from several computer simulations of the sort shown in Fig. 2(c). The results show that the total amount of biomass produced is constant, in the case illustrated 9 units, and that the contribution from each species to this total depends on the inoculum sizes.

to achieve in an industrial or laboratory fermentation system. By analogizing chemostat cultures to open ecosystems microbial ecologists have used Equations (6) to (8) to support Gause's principle of competitive exclusion, which states that no two species can stably occupy the same niche, that is, one of two competing species will eventually be eliminated. However, although coexistence of two species interacting in the way defined by these equations might in nature be rare to the point of nonexistence, it should be possible to maintain such am interaction by manipulating the parameters controlling growth in a chemostat culture. Figure 4 shows simulation results for such a two species system in which the dilution rate has been changed in the manner indicated. Maintaining the *average* density of each population at some desired value could be achieved by monitoring the population densities and adjusting the dilution rate appropriately. On-line computer control would be the obvious choice for such an operation.

De Freitas and Fredrickson [1978] investigated the effect of inhibition on growth rate competition for a limiting nutrient in chemostat culture. The kinetic functions they analysed were of the form:

$$\alpha = \frac{\alpha_i S}{(K_i + S)(1 + I/K_i')} \tag{13}$$

where I is the concentration of inhibitor and K_i' represents an inhibition constant. They found that if

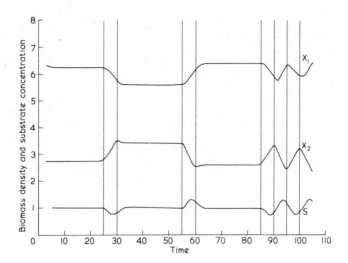

Fig. 4. Simulation of a chemostat culture in which two species are competing with kinetics of the type illustrated in Fig. 2(c). The vertical lines represent changes in dilution rate. The dilution rate at which both species survive is defined by Equation (12).

at least one of the species present produced an autoinhibitor, stable coexistence could occur. This was not the case when one or both of the species produced a substance inhibitory to the other or when a common autoinhibitor was formed. These results open up the possibility of maintaining a two-species system, which is potentially unstable due to competition, by adding growth inhibitors in such a manner as to mimic a growth-linked product.

Competition between higher organisms can be for territory and for factors other than food. By analogy, in the microbial world some organisms appear to compete with each other for room on the surface of solid substrates to which they adhere. McLaren and Ardakani [1972] analysed a model which accounted for competition of this sort between the nitrifying bacteria, *Nitrosomonas* and *Nitrobacter*, growing in a column of soil. With *Nitrosomonas* at density X_1 and *Nitrobacter* at density X_2, the following equations were used:

$$\frac{dX_1}{dt} = (\alpha_1 - \frac{X_1 + k_1 X_2}{K_1}) X_1 \qquad (14)$$

$$\frac{dX_2}{dt} = (\alpha_2 - \frac{X_2 + k_2 X_1}{K_2}) X_2 \qquad (15)$$

The constants K_1 and K_2 represent the maximum population densities that can be attained by the ammonium and nitrite oxidizers while k_1 and k_2 reflect the effect of interspecific competition for sites on a solid surface. Equations (14) and (15) are special cases of Equations (1) and (2) but their interpretation by McLaren and Ardakani [1972] with respect to competition differs from that of Fredrickson [1977]. The latter author regards competition as an indirect interaction. In the nitrification system competition between two species for the physical occupation of a microenvironment occurs and contact between the species is likely to occur. This type of competition then is a direct interaction involving the β_{ij} functions of Equations (1) and (2) such that:

$$\beta_{12} = k_1/K_1$$

and

$$\beta_{21} = k_2/K_2$$

4. Commensalism and Mutualism

In these associations one or both of the populations benefit from the presence of the other. In kinetic terms, growth-promoting substances are assumed to be produced which increase the specific growth rates of the populations. Thus in the case of commensalism, in which only one species benefits, α_1 might be an increasing function of P_2, the concentration of a product manufactured by the second species present. For example, in terms of Monod kinetics:

$$\alpha_1 = \frac{\alpha_{m1} P_2}{K_1 + P_2} \qquad (16)$$

while for the second species:

$$\alpha_2 = \frac{\alpha_{m2} S}{K_2 + S} \qquad (17)$$

where S is the concentration of some externally supplied limiting nutrient. In the case of mutualism both species benefit and so in terms of soluble growth-promoting products and Monod kinetics, Equation (16) holds and Equation (17) would take on a similar form, that is:

$$\alpha_2 = \frac{\alpha_{m2} P_1}{K_2 + P_1} \qquad (18)$$

where P_1 is the concentration of a product made by the first population.

It is, of course, possible for more than one interaction to occur between two species, and in many situations this is frequently the case. An interesting association of this sort was investigated by Megee et al. [1972]. These workers grew *Saccharomyces cerevisae* and *Lactobacillus casei* together in chemostat culture. The yeast produced riboflavin necessary for the growth of the bacterium and the limiting carbon source in the system was glucose. The kinetic model employed by the authors included a double substrate limitation term for *L.casei* of the form:

$$\alpha_1 = \frac{\alpha_{m1} S P_2}{(K_1+S)(K'_1+P_2)} \quad (19)$$

where S is the concentration of glucose and P_2 is the concentration of riboflavin. The specific growth rate of the yeast was assumed to be a function of glucose only:

$$\alpha_1 = \frac{\alpha_{m2} S}{K_2+S} \quad (20)$$

A maintenance term was included in the equations representing the rate of change in the yeast population. Their experiments and their simulations gave similar results. When excess riboflavin was added to the feed medium, competition occurred between the two species and only the bacteria survived because they grew faster than the yeast. The effect of competition was therefore to destabilize the ecosystem. Limiting the amount of riboflavin in the medium resulted in a mutualistic relationship in which the riboflavin produced by the yeast was used in addition to that supplied in the feed. The relatively high densities of bacteria produced sufficient lactic acid to reduce the pH of the medium enough to significantly increase the specific growth rate of the yeast. When no riboflavin was present in the input medium a commensal relationship between the two populations occurred with the sole source of riboflavin being that produced by the yeast which effectively limited the population density of the bacterium and lactic acid production. In both the commensal and mutualistic association the limiting carbon source was glucose. The effect of these associations was therefore to stop one of the species becoming extinct by competitive exclusion. Thus mutualism and commensalism stabilized the ecosystem. Competition, on the other hand, destabilized the system. In fact it must be rare that competition ever does stabilize an ecosystem contrary to Slater and Bull's [1978] contention that the stability of the *Lactobacillus/Saccharomyces* system was a result of competition.

5. Predation and Parasitism

In kinetic terms the models used to study predator-prey relationships are often identical to those used to study the interactions between host and parasite populations. For this reason I will talk only in terms of predator-prey dynamics.

The interaction between prey and predator populations is direct and so the β_{ij} functions in Equations (1) and (2) are non-zero. With $\beta_{11} = \beta_{22} = 0$ these equations can simplify to

$$\frac{dX_1}{dt} = \alpha_1 X_1 - \beta_{12} X_1 X_2 \qquad (21)$$

$$\frac{dX_2}{dt} = -\alpha_2 X_2 + \beta_{21} X_1 X_2 \qquad (22)$$

where all the coefficients are constants and X_1 represents the density of the prey population and X_2 the density of the predator. These equations imply that the prey population increases exponentially without limit in the absence of predators, predation is obligate with the predator dying exponentially in the absence of prey, a direct proportionality between predator specific growth rate and prey density exists and that the increase in predator density brought about by predation is proportional to the decrease in prey density.

By dividing Equation (21) by Equation (22) an expression relating X_1 to X_2 can be obtained:

$$\alpha_1 \log X_2 - \beta_{12} X_2 + \alpha_2 \log X_1 - \beta_{21} X_1 = C \qquad (23)$$

where C is a constant of integration numerically equal to the value of the left-hand side of Equation (23) with X_1 and X_2 at their values at time t = 0. Solutions of Equation (23), depicted in Fig. 5, are a set of closed trajectories, each trajectory dependent on the initial conditions of the system. The point about which the closed curves move is called the equilibrium or stationary point and is defined by the values that X_1 and X_2 take on when $dX_1/dt = dX_2/dt = 0$. The direction of a trajectory is found by determining the behaviour of the derivatives with respect to the equilibrium point. The broken lines in Fig. 5 indicate zero values for the two derivatives and subdivide the plane into quadrants denoted A — D. In each quadrant the sign of dX_1/dt and dX_2/dt can be assessed by inspection. The sign of dX_2/dX_1 can therefore be determined. The slope of the trajectory is dX_2/dX_1 so if this derivative is positive the arrow points upwards and if it is negative the arrow points downwards. In quadrant A, for example, $dX_2/dt > 0$ and $dX_2/dt > 0$ so $dX_2/dX_1 > 0$ and the arrow points upwards. In quadrant B, $dX_2/dt > 0$, $dX_1/dt < 0$ so $dX_2/dX_1 < 0$ and the motion of the trajectory is downwards.

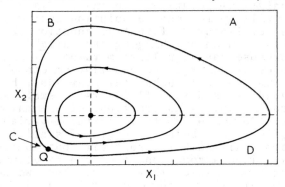

Fig. 5. Solution of Equation (23) for populations of a prey, X_1 and predator, X_2, in predator-prey phase space. Each closed trajectory represents the results obtained with a different value of the initial conditions-dependent constant, C.

Figure 5 can be used to determine qualitatively the behaviour of the predator and prey populations with respect to time once the direction of the trajectories is known. Consider the outer trajectory at the point Q in quandrant C. As the prey population (X_1) increases so the predator (X_2) declines slightly, then increases in quadrant D, slowly at first and then rapidly. The prey reaches its maximum density in quadrant D and then declines rapidly in quadrant A while the predator population increases to its maximum. In quadrant B both populations decline, the predator at an increasing rate and the prey at a decreasing rate. This behaviour implies sustained, out-of-phase oscillations in the population densities of the two interacting species. The amplitude of the oscillations depends on which particular trajectory the system is in and so they are dependent on the initial conditions. This means that if the system is perturbed the amplitudes change. In this respect the system is unstable.

The discussion above indicates how important are the shapes of the closed curves in phase space generated by the Lotka-Volterra equations. An important feature is that the shapes of the curves change with respect to their distance from the equilibrium point. Near the equilibrium point they resemble ellipses and Pielou [1969] uses this feature to derive expressions for the time-dependent behaviour of the prey and predator populations near equilibrium. A more conventional approach is to use Liapounov stability analysis [Liapounov, 1947].

I reproduce part of such an analysis below to derive the essential elements of the results.

Equations (21) and (22) constitute a set of linked, non-linear differential equations for which we do not have an analytical solution. In the analysis which follows we will approximate a solution near equilibrium using a set of linear differential equations which we can solve. This is done by changing the variables X_1 and X_2, which we suppose have equilibrium values $\tilde{X}_1$ and $\tilde{X}_2$, to new variables x_1 and x_2, the equilibrium coordinates of which are 0,0. As only behaviour near equilibrium will be considered, x_1 and x_2 will always be small numbers which we could express as fractions of the maximum population sizes. Second order terms in these variables, that is x_1^2, x_2^2 or $x_1 x_2$, being products of small fractions, will be very small numbers, so their contribution need not be taken into account and they can be eliminated from the expressions in which they occur. When we do this we find that we are left with a set of linear differential equations with a well-known solution.

We define:
$$x_1 = X_1 - \tilde{X}_1$$
and
$$x_2 = X_2 - \tilde{X}_2$$

Substituting for X_1 and X_2 in Equations (21) and (22) yields:

$$\frac{dx_1}{dt} = \alpha_1(x_1+\tilde{X}_1) - \beta_{12}(x_1+\tilde{X}_1)(x_2+\tilde{X}_2) \qquad (24)$$

$$\frac{dx_2}{dt} = -\alpha_2(x_2+\tilde{X}_2) + \beta_{21}(x_1+\tilde{X}_1)(x_2+\tilde{X}_2) \qquad (25)$$

Multiplying out Equation (24) and rearranging terms gives:

$$\frac{dx_1}{dt} = \alpha_1\tilde{X}_1 - \beta_{12}\tilde{X}_1\tilde{X}_2 + \alpha_1 x_1 - \beta_{12}\tilde{X}_2 x_1 \qquad (26)$$
$$- \beta_{12} x_1 x_2 - \beta_{12}\tilde{X}_1 x_2$$

From Equation (21) and by definition:
$$\frac{dx_1}{dt} = 0 = \alpha_1 \tilde{X}_1 - \beta_{12}\tilde{X}_1\tilde{X}_2 \qquad (27)$$

so the first two terms in Equation (26) can be eliminated. Similarly, the third and fourth terms equate to zero. The fifth term is second order in x_1 and x_2 so it represents a very small number and can be disregarded. Therefore we are left only with:

$$\frac{dx_1}{dt} = -\beta_{12}\tilde{X}_1 x_2 \qquad (28)$$

By a similar process we can derive from Equation (25)

$$\frac{dx_2}{dt} = \beta_{21}\tilde{X}_2 x_1 \qquad (29)$$

We now have a set of two simple differential equations for x_1 and x_2 but they are still non-linear. We obtain linear equations as follows. Differentiation of Equation (28) yields:

$$\frac{d^2 x_1}{dt^2} = -\beta_{12}\tilde{X}_1 \frac{dx_2}{dt} \qquad (30)$$

Using Equation (29) we can eliminate terms in x_2 by substitution to get a second order, linear differential equation in x_1.

$$\frac{d^2 x_1}{dt^2} = -\beta_{12}\beta_{21}\tilde{X}_1\tilde{X}_2 x_1 \qquad (31)$$

This equation can be simplified further by replacing the equilibrium values with:

$$\tilde{X}_1 = \alpha_2/\beta_{21}$$

and

$$\tilde{X}_2 = \alpha_1 \beta_{12}$$

and by rearranging we get:

$$\frac{d^2 x_1}{dt^2} + \alpha_1 \alpha_2 x_1 = 0 \qquad (32)$$

Equation (32) is the equation for simple harmonic motion and has the oscillatory solution:

$$x_1 = A \cos(\omega t + \delta) \qquad (33)$$

where A is the amplitude of the oscillations and is a constant of integration (which depends on the initial conditions), ω is the angular frequency which determines the periodicity of the oscillations and δ is the phase constant. If we assume that at time $t = 0$, x_1 is at its maximum value then $\delta = 0$.

The equations for the predator population near equilibrium give a similar result except that a sine function is generated

$$x_2 = A' \sin(\omega t + \delta') \qquad (34)$$

From Equations (33) and (34) we see that according to the Lotka-Volterra model, near equilibrium the prey and predator population will oscillate continuously one quarter phase out of step with each other, with the prey increasing first because it changes according to a cosine function while the predator density is described by a sine function. The periodicity (T) of both populations is the same and depends on the growth characteristics of the species involved because:

$$T = \frac{2\pi}{\omega} \qquad (35)$$

and

$$\omega = \sqrt{\alpha_1 \alpha_2} \qquad (36)$$

On the other hand, the amplitudes of the oscillations, A and A', are dependent on the initial conditions and change after environmental perturbations. Figure 6 gives an example of simple Lotka-Volterra behaviour.

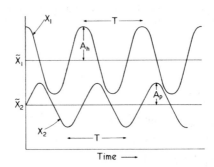

Fig. 6. Predator-prey dynamics according to the Lotka-Volterra equations where X_1 is prey density and X_2 is predator density. A_h is the amplitude of the prey fluctuation and A_p that of the predator. T is the amplitude of the oscillation and $\tilde{X}_1$ and $\tilde{X}_2$ the equilibrium value of the populations.

In microbiology the Monod function is often used to represent the dynamic effect of predation. If a similar term is used to describe prey growth then a third equation is needed for prey substrate and the resulting set of equations represents a food chain in which a limiting nutrient, S, is converted to prey biomass, X_1, which, in turn, is converted to predator, X_2. The equations for predation in a chemostat are:

$$\frac{dS}{dt} = D(S_r - S) - \frac{\alpha_1 S X_1}{Y(K_1 + S)} \qquad (37)$$

$$\frac{dX_1}{dt} = \frac{\alpha_1 S X_1}{K_1 + S} - \frac{\beta_1 X_1 X_2}{K_2 + X_1} - DX_1 \qquad (38)$$

$$\frac{dX_2}{dt} = \frac{\beta_2 X_1 X_2}{K_2 + X_1} - DX_2 \qquad (39)$$

where: α_1, K_1 and Y are the maximum specific growth rate, saturation constant and growth yield of the prey population; β_2, K_2 and Z are the same quantities for the predator population such that $\beta_2 = \beta_1/Z$; and S_r is the concentration of limiting nutrient entering the system.

It is apparent that the mathematical model of microbial predation represented by Equations (37) to (39) cannot be described readily using the Lotka-Volterra framework of Equations (1) and (2). By comparing Equations (1) and (2) we could let:

$$\alpha_1 = \frac{\alpha_1 S}{K_1 + S} \qquad (40)$$

$$\beta_{12} = \frac{\beta_1}{Z(K_2 + X_1)} \qquad (41)$$

and

$$\beta_{11} = \frac{D}{X_1} \qquad (42)$$

although the relationship between β_{11} and D is rather artificial. When we relate Equation (2) to Equation (37), however, problems of a more serious nature arise. If we let:

$$\beta_{21} = \frac{\beta_2}{K + X} \quad \text{with } \alpha_2 = 0 \qquad (43)$$

the predator specific growth rate function is zero despite the fact that we have used a Monod term, conventionally associated with the specific growth rate of a microbial population, to describe the way the predator population grows. Alternatively, if:

$$\alpha_2 = \frac{\beta_2 X_1}{K_2 + K_1} \quad \text{and} \quad \beta_{21} = 0 \tag{44}$$

an asymmetry is imposed in that the term representing a direct interaction in the prey equation becomes a specific growth rate function in the predator equation. Neither alternative is readily acceptable which illustrates the point that although formulation of microbial interactions in terms of Equations (1) and (2) is useful for purposes of definitions, it has limited value as a basis for classifying the kinetic functions which have been used to describe interaction dynamics.

Equations (37) to (39) can be analysed by the Liapounov method [Canale, 1970] by which means it can be shown that three qualitatively different types of nontrivial solution can be obtained, that is, a solution by which both prey and predator populations survive. The responses predicted depend on the dilution rate employed and are sketched diagrammatically in Fig. 7. At relatively high dilution rates, both predator and prey populations move smoothly and monotonically to steady state densities. At intermediate dilution rates, damped oscillations occur and at low values of D sustained oscillations occur. The amplitudes of the undamped limit cycle oscillations are not dependent on the initial conditions and perturbing the system only transiently affects them. They are thus quite different from the sinusoidal fluctuations generated by the Lotka-Volterra equations.

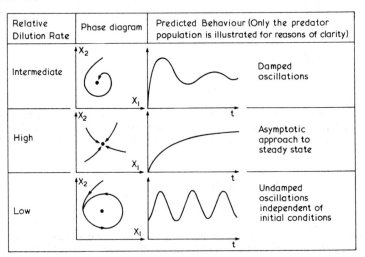

Fig. 7. Diagrammatic representation of the response of the predator-prey system defined by Equations (37) − (39).

Of the three types of response predicted by Monod kinetics for microbial predation in chemostat culture damped oscillations have been observed experimentally [Tsuchiya et al., 1972; Canale et al., 1973; Dent et al., 1976]. Results of this sort for slime mould amoebae feeding on *Escherichia coli* are shown in Fig. 8. However, as with the data of Tsuchiya et al. [1972] and Dent et al. [1976], these results differ quantitatively from those predicted by Monod kinetics. Thus, although there is some degree of qualitative agreement between theoretical prediction and experimental results, there is not quantitative agreement.

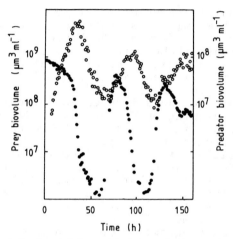

Fig. 8. Growth of *Escherichia coli* (closed circles) and *Dictyostelium discoideum* amoebae (open circles) in chemostat culture.

In earlier work of Curds and Cockburn [1968], batch culture results were obtained in which predation by *Tetrahymena pyriformis* on *Aerobacter aerogenes* seemed to correspond to a kinetic model proposed by Contois [1959] for bacteria growing on a soluble substrate. When this expression is incorporated, Equations (38) and (39) become:

$$\frac{dX_1}{dt} = \frac{\alpha_1 S X_1}{K_1 + S} - \frac{\beta_1 X_1 X_2}{K_2 X_2 + X_1} - DX \tag{45}$$

$$\frac{dX_2}{dt} = \frac{\beta_2 X_1 X_2}{K_2 X_2 + X_1} - DX_2 \tag{46}$$

Under steady state conditions, then:

$$D = \frac{\beta_2 \tilde{X}_1}{K_2 \tilde{X}_2 + \tilde{X}_1} \quad (47)$$

which can be rearranged to:

$$D = \frac{\beta_2 (\tilde{X}_1/\tilde{X}_2)}{K_2 + (\tilde{X}_1/\tilde{X}_2)} \quad (48)$$

Here D is equal to the specific rate of growth of the predator which we will designate λ. Equation (48) has the same functional form as the Monod expression, that is, a rectangular hyperbola, but in the latter case the independent variable in terms of predation kinetics is prey density (X_1) whereas in the equation it is the ratio of prey to predator (X_1/X_2).

Owen [1979] used data from chemostat cultures of slime mould amoebae growing on *Escherichia coli* (of the sort shown in Fig. 8), smoothed it by the method of cubic splines and calculated the specific rate of change of the amoebal population, that is λ, from the slope of the line generated. When λ was plotted against the ratio of prey to predator densities, what appeared to be a family of Monod-type saturation curves was produced with both the saturation constant and maximum specific growth rate decreasing with time. This result therefore supports the idea that predator growth depends on the ratio of prey to predator and in addition suggests that the interaction function β_{ij} is time-dependent.

The dependence of λ on the amount of prey available to each predator and also on time was supported by the analysis of the slime mould/*Escherichia coli* chemostat system by Bazin and Saunders [1978] using catastrophe theory. If the functions of Equations (1) and (2) are indeed time-dependent, then the interaction becomes considerably more complicated than previously supposed. Equations (1) and (2) are usually taken to be autonomous in that time does not appear on the right-hand side of the equals sign. If the equations are non-autonomous, a far greater variety of dynamic behaviour is possible.

6. Multispecies Systems

We have seen that the Lotka-Volterra formulation of equations (1) and (2) is useful for defining the ways in which microbial species interact with each other and for illustrating how dynamic equations can be analysed. They have also been of value in the theory of multispecies interaction. In particular, they have been used to clarify, at least to some extent, the relationship between stability, diversity and complexity of a multi-species system.

Up until about 1970 it was supposed by ecologists that ecosystems became more stable as the diversity of species they contained increased. Several measures of diversity were devised which reflected at the same time both the number of species present and the number of individuals. One measure of diversity is the so-called Shannon diversity index,

$$\bar{H} = -k \sum_{i=1}^{m} p_i \log p_i \qquad (49)$$

where p_i is the probability of selecting a species of the ith type in a community containing m species. A derivation of this relationship is given by Patten [1962]. As May [1976] points out, this diversity index "is linked by an ectoplasmic thread to information theory", but because of this link it has been assumed that, like other connected systems such as telephone networks, the number of connections between species in an ecological community is a reflection of the community's stability. That is to say, the more interactions there are between species, the more stable will be the system as a whole. This proposition has been investigated using the Lotka-Volterra equations. In order to understand the methods used it is necessary to look in a little more detail at the behaviour of these equations near equilibrium and how they are solved.

In the section on predation I used a particular form of the Lotka-Volterra equation which when analysed near equilibrium gave Equation (32). This is a second order equation in terms of x_1, the variable representing the population X_1 near equilibrium. For the more general Lotka-Volterra equations (1) and (2), the linearized equations analogous to Equations (28) and (29) are:

$$\frac{dx_1}{dt} = \beta_{12}\tilde{X}_1 x_2 + \beta_{11}\tilde{X}_1 x_1 \qquad (50)$$

and

$$\frac{dx_2}{dt} = \beta_{21}\tilde{X}_2 x_1 + \beta_{22}\tilde{X}_2 x_2 \qquad (51)$$

while Equation (32) becomes:

$$a\frac{d^2 x_1}{dt^2} + b\frac{dx_1}{dt} + cx_1 = 0 \qquad (52)$$

with a = 1

$$b = -(\beta_{11}\tilde{X}_1 + \beta_{22}\tilde{X}_2)$$

$$c = (\beta_{11}\beta_{22} - \beta_{12}\beta_{21})\tilde{X}_1\tilde{X}_2$$

where, as before, a tilde over a variable represents an equilibrium value. Note that Equation (52) has the same form as Equation (32) but in the latter case b = 0. Now the general solution of Equation (52) is

$$x_1 = k_1 e^{\gamma_1 t} + k_2 e^{\gamma_2 t} \qquad (53)$$

where k_1 and k_2 are constants of integration dependent on the initial condition of the system. The exponents γ_1 and γ_2 are eigenvalues obtained from the coefficients in Equation (52) by solving the quadratic equation:

$$a\gamma^2 + b\gamma + c = 0 \qquad (54)$$

This is done using the quadratic formula so that:

$$\gamma = \frac{-b \pm \sqrt{b^2 - 4ac}}{2a} \qquad (55)$$

By solving Equation (55) we get two values for γ which can be substituted into Equation (53) to determine how x_1 behaves with respect to time. For the curves where both values of γ are positive, x_1 increases without limit in time and the system is said to be unstable. When γ_1 and γ_2 are both negative, x_1 tends towards zero with respect to time so X_1 tends towards its equilibrium value, $\tilde{X}_1$. In this case the system is stable. For γ_1 and γ_2 of opposite sign the system is unstable for although the system might transiently move towards its equilibrium, this condition cannot be sustained.

At this point I will digress from the chief direction of this section to explain how the sinusoidal solution in Equation (33) resulted from Equation (32). In the solution of Equation (55) to determine the eigenvalue for the system, sometimes the term 4ac is bigger than b. In such cases it is necessary to determine the square root of a negative number. This is accomplished by defining i as the square root of -1 and writing the solution of Equation (55) in the form:

$$\gamma = n \pm pi \qquad (56)$$

which is called a complex conjugate pair in which n is a real number and pi is an imaginary number. Equation (53) therefore becomes:

$$x_1 = k_1 e^{(n+pi)t} + k_2 e^{(n-pi)t} \tag{57}$$

Now it can be shown that an exponential raised to an imaginary number can be represented by a sine (or cosine) function. Thus the imaginary part of the complex conjugate determines the oscillatory nature of the solution and the real part determines whether it is stable or not; positive values of n cause the oscillation to increase in magnitude and the system is unstable. Negative values of n result in damped oscillations and the system is stable. Applying these arguments to Equation (32) we find that:

$$\gamma = \pm \sqrt{-\alpha_1 \alpha_2} \tag{58}$$

$$= \pm \sqrt{\alpha_1 \alpha_2}\, i$$

That is, in this special case n = 0 so that the oscillations in x_1 neither increase nor decrease in magnitude and an undamped sinusoid of the form in Equation (33) results.

To return to the question of ecosystem stability, we have shown that this property depends on whether the eigenvalues are negative or not. For a two-species system there are two eigenvalues while for an m-species system there are m eigenvalues, all of which have to be negative for the system to be stable. For a multi-species system Equations (1) and (2) are conveniently written as:

$$\dot{X}_i = \alpha_i X_i + \sum_{i=1}^{m} \beta_{ij} X_i X_j \tag{59}$$

where the dot over a variable represents its derivative with respect to time. Using matrix notation the linearized Equations (50) and (51) for a two-species system can be written as:

$$\begin{bmatrix} \dot{x}_i \\ \dot{x}_2 \end{bmatrix} = \begin{bmatrix} \beta_{11}\tilde{X}_1 & \beta_{12}\tilde{X}_1 \\ \beta_{21}\tilde{X}_2 & \beta_{22}\tilde{X}_2 \end{bmatrix} \begin{bmatrix} x_1 \\ x_2 \end{bmatrix} \tag{60}$$

and the characteristic equation equivalent to Equation (54) using a determinant is:

$$\begin{vmatrix} \beta_{11}\tilde{X}_1 - \gamma & \beta_{12}\tilde{X}_1 \\ \beta_{21}\tilde{X}_2 & \beta_{22}\tilde{X}_2 - \gamma \end{vmatrix} = 0 \qquad (61)$$

an m-species system can be represented in the same way:

$$\begin{bmatrix} \dot{x}_1 \\ \dot{x}_2 \\ \dot{x}_3 \\ \vdots \\ \dot{x}_m \end{bmatrix} = \begin{bmatrix} a_{11} & a_{12} & \cdots & a_{1m} \\ a_{21} & a_{22} & & \vdots \\ a_{31} & a_{32} & a & \vdots \\ \vdots & & & \vdots \\ a_{m1} & \cdots & & a_{mm} \end{bmatrix} \begin{bmatrix} x_1 \\ x_2 \\ x_3 \\ \vdots \\ x_m \end{bmatrix} \qquad (62)$$

with the characteristic equation:

$$\begin{vmatrix} a_{11} - \gamma & a_{12} & \cdots & a_{1m} \\ a_{21} & a_{22} - \gamma & \cdots & \vdots \\ a_{31} & a_{32} & a_{33} - \gamma & \cdots \\ \vdots & & & \vdots \\ a_{m1} & \cdots & & a_{mm} - \gamma \end{vmatrix} = 0 \qquad (63)$$

In the last two equations the entries, a_{ij}, represent combinations of the coefficients of the original equations, i.e., combinations of the α_i's and β_{ij}'s. The solution of Equation (63) will give numerical values for each of the eigenvalues involved. But in order to determine the stability of the system all we need to know is whether or not the eigenvalues are negative. To determine this we need not perform the entire calculation necessary for evaluating the eigenvalues; all that is necessary is application of what are called the Routh-Hurwitz criteria, which indicate whether or not the eigenvalues are negative and thereby the stability of the system.

In practice stability analysis is usually performed by using a computer library program so that the stability

of the system can be determined simply by supplying the
values for the m x m matrix in Equation (63). As I
mentioned before, these entries represent combinations of
the coefficients in the dynamic equations. A minimum
requirement of stability is that all the diagonal elements
of the matrix be negative. The off-diagonal elements
represent interactions. Zero non-diagonal elements represent no interactions so the number of non-zero off-diagonal elements represents the number of interactions
in the system or connections. Gardner and Ashby [1970]
investigated linearly connected systems of the type represented by Equation (63) and their results can be interpreted in terms of the proposition introduced at the
beginning of this section, that is to say, does stability
increase with the number of connections in a system?
What Gardner and Ashby [1970] did was to construct
matrices representing ecosystems. They gave all the
diagonal elements negative values, thus assuming that
a priori the matrices were not unstable. Then they
specified a degree of connectance, C, defined as the
fraction of non-zero, off-diagonal entries and assigned
values between -1 and +1 to C% of the off-diagonal
elements. They then applied the Routh-Hurwitz criteria
to determine stability. By repeating this process
several times they were able to estimate the probability
of stability for a matrix of given size and connectance.
They found that the chance of a system being stable
decreased with the value of C. In terms of a multi-species system, then, stability decreases as the number
of interactions increases. This result is just the
opposite of that supposed by ecologists but of course
depends on the assumption that the interactions between
the species involved can be represented in a linear
fashion. The most plausible interpretation of these
theoretical results is that this is not the case and that
in fact the kinetic functions describing species interactions are non-linear. This analysis in terms of the
linearized Lotka-Volterra equation shows that stability
is not a direct result of the complexity of connected
systems and therefore the kinetics of the interactions
must be of such a nature as to give rise to mixed
populations that are stable.

7. Conclusions

I have tried in this chapter to illustrate what I
consider to be some of the more interesting current
approaches to the study of microbial interaction kinetics.
I have shown, hopefully, that although formulation of
mixed culture systems in terms of the Lotka-Volterra
equations does not represent the dynamics of microbial
interactions in a realistic way and does not form a
suitable basis for the classification of the types of
interactions between species, they are useful in defining

terms and in pointing the route along which future research into multispecies systems might take. For a given multispecies community it might be possible one day in the future to write down a set of differential equations which describe accurately the dynamics of the system. But at the moment we cannot do this for most two-species interactions and in any case is a complete description of this sort really an appropriate goal for this type of research? The probability in many cases is that at least some of the most important characteristics of multispecies systems can be uncovered by methods similar to the analysis of stability I have illustrated using the Lotka-Volterra equations. It is upon such an assumption that this chapter was based.

References

Bazin, M.J. and Saunders, P.T. (1978). Determination of critical variables in a microbial predator-prey system by catastrophe theory. *Nature* **275**, 52-54.

Canale, R.P. (1970). An analysis of models describing predator-prey interaction. *Biotechnology and Bioengineering* **12**, 353-378.

Canale, R.P., Lustig, T.D., Kehrberger, P.M. and Salo, J.E. (1973). Experimental and mathematical modelling studies of protozoan predation on bacteria. *Biotechnology and Bioengineering* **15**, 707-728.

Contois, D.E. (1959). Kinetics of bacterial growth: Relationship between population density and specific growth rate of continuous cultures. *Journal of General Microbiology* **21**, 40-50.

Curds, C.R. and Cockburn, A. (1968). Studies on the growth and feeding of *Tetrahymena pyriformis* in axenic and monoxenic culture. *Journal of General Microbiology* **54**, 343-358.

De Freitas, M.J. and Fredrickson, A.G. (1978). Inhibition as a factor in the maintenance of the diversity of microbial ecosystems. *Journal of General Microbiology* **106**, 307-320.

Dent, V.E., Bazin, M.J. and Saunders, P.T. (1976). Behaviour of *Dictyostelium discoideum* amoebae and *Escherichia coli* grown together in chemostat culture. *Archives of Microbiology* **109**, 187-194.

Fredrickson, A.G. (1977). Behaviour of mixed cultures of microorganisms. *Annual Review of Microbiology* **31**, 63-87.

Gardner, M. and Ashby, W. (1970). Connectance of large dynamic (cybeonetic) systems: Critical values of stability. *Nature, London* **228**, 784.

Harrison, D.E.F. (1978). Mixed cultures in industrial fermentation processes. *Advances in Applied Microbiology* **24**, 129-164.

Liapounov, M.A. (1947). Problème génerale de la stabilité du mouvement. Princeton: Princeton University Press.

May, R.M. (1976). Patterns in multispecies communities. In "Theoretical Ecology: Principles and Applications", (ed. R.M. May), pp.142-162. Oxford: Blackwell Scientific Publications.

McLaren, A.D. and Ardakani, M.S. (1972). Competition between species during nitrification in soil. *Proceedings of the Soil Science Society of America* **36**, 602-606.

MeGee, R.D., Drake, J.F., Fredrickson, A.G. and Tsuchiya, H.M. (1972). Studies in intermicrobial symbiosis. *Saccharomyces cerevesiae* and *Lactobacillus casei*. *Canadian Journal of Microbiology* **18**, 1333-1742.

Owen, B.A. (1979). Growth of *Dictyostelium discoideum* amoebae in chemostat culture. Ph.D. thesis, University of London.

Patten, B.C. (1962). Species diversity in net phytoplankton of Raritan Bay. *Journal of Marine Research* **20**, 57-75.

Pielou, E.C. (1969). "An Introduction to Mathematical Ecology". New York: Wiley Interscience.

Slater, J.H. (1979). Microbial population and community dynamics. In "Microbial Ecology: A Conceptual Approach", (eds. J.M. Lynch and N.J. Poole), pp.45-63. Oxford: Blackwell Scientific Publications.

Slater, J.H. and Bull, A.T. (1978). Interactions between microbial populations. In "Companion to Microbiology", (eds. A.T. Bull and P.M. Meadow), pp.181-206. London and New York: Longman.

Tsuchiya, H.M., Drake, J.F., Jost, J.L. and Fredrickson, A.G. (1972). Predator-prey interactions of *Dictyostelium discoideum* and *Eschericgia coli* in continuous culture. *Journal of Bacteriology* **110**, 1147-1153.

Chapter 3

MICROBIAL PATHWAYS AND INTERACTIONS
IN THE ANAEROBIC TREATMENT PROCESS

PETER N. HOBSON

*The Rowett Research Institute, Greenburn Road,
Bucksburn, Aberdeen, Scotland, UK*

1. Introduction

Anaerobic digestion is a response to controlled conditions of a reaction, or more properly a series of reactions, that occurs in very many circumstances in nature. This is the conversion of "dead" organic matter to the gases methane and carbon dioxide and biomass. The term "dead" matter is used here since, while living organic materials can be attacked by microorganisms, the fact that the vegetation is living means that it is surrounded by air and the microbes have to work largely in the presence of oxygen, while the various stages in the production of gas are essentially fermentative and have to take place in the absence of oxygen and at very low redox potential. So, the reactions take place in nature under still water or wet soil or in similar places where the ingress of oxygen from the air is limited or precluded, and we find evidence in the gas bubbles arising from stagnant pools and marshes, from mud-flats in rivers and from piles of rotting vegetation in rubbish dumps.

The frequency with which this reaction occurs and results of tests on rumen anaerobic bacteria show that although the bacteria require highly-reduced conditions for growth, they can survive exposure to air when not metabolizing and can be transferred by air or water to colonize new habitats [Mann, 1963]. Although the rumen bacteria may not grow, because of rate of passage of digesta or other factors, they are picked up from the ground or air by animals and pass through the digestive tract and are found in faeces. These bacteria in faecal material or in the air or earth form the inoculum for the industrial process of anaerobic digestion. It is a process using an undefined, and largely unknown, mixture of bacteria which comes from a natural inoculum but which settles down to a stable mixture of species breaking down the particular feedstock used. The system is self-regulating in the sense that, as we shall discuss later, there are few "mechanical" constraints put on the process. This does not, however, mean that once inoculated and stabilized the same bacteria are present *ad infinitum*. Since process conditions are non-sterile, and in most or all cases the bacteria are present in the feedstock, a continuous variation in strain or type of bacteria almost certainly occurs as it does in the rumen [Hobson *et al*., 1958], even if the species remain the same.

The overall process is similar to that occurring in the rumen. Experiments have shown that it is extremely difficult or, in our present stage of knowledge, with some feeds impossible to reproduce adequately the rumen function with a defined flora in gnotobiotic lambs [Lysons *et al*., 1971, 1976; Mann and Stewart, 1974]. So it would seem equally impossible to reproduce the digester functions with defined microbial floras. In addition, tests have shown that the defined rumen flora, even if apparently sufficient for feed degradation and metabolism and with bacteria present in high numbers, rapidly reverts to the natural mixture when the host lambs are removed from their sterile environment. Thus, a defined digester flora would, presumably, revert to the natural flora under normal digester operating conditions. However, although a defined digester flora may not be practicable, a better understanding of the bacteria and the reactions involved in the process can help in improving digester design and operation and in assessing different types of digesters.

2. Digester Feedstocks and Design

While the relative rates of reactions and the growth rates of the bacteria, as well as optimum conditions for reactions, can be considered in various ways in designing digester systems, as will be shown later, much of this is at present theoretical or at the laboratory-scale stage. What is in practice possible is governed by the physical nature of the feedstock and by the fact, initially deduced from running of digesters and now being confirmed by laboratory studies, that many of the reactions are slow.

Digestion in its oldest and widest application is part of the municipal sewage treatment process and is an extension of the septic tank once used for town sewage, but now mainly confined to small (one or two house) units. Water usage has increased over the years and municipal sewage is now a very large volume of dilute aqueous solution with a low suspended solids content. The suspended solids, however, consist of a wide range of particle sizes from colloidal and microscopic up to large pieces of paper and other debris. The heavier particles, after preliminary screening, are settled out of the sewage as it enters the sewage works and removed as primary sludges for some form of treatment. The sewage water then passes on to an aerobic treatment, generally a type of activated sludge system. The aerobic treatment generates more sludge of gelatinous, aerobic, microorganisms plus fine debris. Some of this may be recycled but excess sludge has to be treated with the preliminary sludge. If it is not treated the sludge rapidly putrifies and causes pollution problems. The method of treating the sludge in many sewage works is by anaerobic digestion.

The primary sludge contains faecal material. This is intestinal bacterial cells, secretions from the gut and food residues. In addition the sludge contains scraps of toilet and other papers, residues from food preparation, detergent residues and debris from street drains. It may also, depending on the local environment, contain waste from factories and the presence of the aerobic sludge has already been mentioned. In addition, the sludges contain biologically inert materials, at least, inert to anaerobic metabolism. These are mineral oils, plastic materials, grit, soil and stones. The broad chemical composition of combined sludges varies according to the proportions of primary and aerobic sludge and with the collecting area, but fats are often a significant proportion of domestic sewage sludges.

The sludge finally fed to the digesters is usually about 4% total solids (TS) although in some systems it may be further thickened to 6 or 8% TS. Although there are some substances in solution, most of the bacterial substrates are contained in the solids suspended in the sludge, indicating that microbial breakdown will be slow. The sludge is produced continuously and so treatment must be continuous, and this is given by the "high-rate" digester. This is a single-stage, stirred-tank, continuous culture system operated at a mesophilic temperature and at a dilution rate measured in reciprocal days (compared with the reciprocal hours of the laboratory or industrial dissolved substrate culture). Generally retention or detention times (i.e., the reciprocal of the dilution rate) rather than dilution rates are quoted and sewage digesters usually have a 20 to 30 d detention time. The consistency of the sludge feedstock, together with its content of grit and other undegradable materials, makes it much more difficult to pump and mix than the dissolved feedstock of the usual industrial fermenter and this results in some digesters operating under non-optimal conditions. For instance, Brade and Noone [1979] quoted a number of investigations of sewage digesters which showed that in a large proportion inadequate mixing of the digester contents led to "dead" areas where microbial activity was slow or nil and that, because of accumulations of grit and undegraded materials, the effective volumes of the digesters were less than the nominal volume. In such circumstances the nominal detention time

was obviously different from the actual detention time.
Sludges require quite large-bore pipes and pumps to ensure
blockage-free passage and since, with the long detention
times used, hourly digester feed-volumes are comparatively
small, intermittent pump operation is generally sufficient
to supply the feed. For this and other reasons most
digesters depart from the theoretical continuous-flow
input characteristics. Indeed, some digesters are fed at
intervals of a few days. This can decrease efficiency,
as was shown for pig-waste digesters fed daily or every
5 min [Bousfield et al., 1974]. Temperature control,
both as regards even temperature throughout the digester
and day-to-day stability, has not always been optimum
[Brade and Noone, 1979] and has led to inefficient
digestion.

The 20 to 30 d detention time is probably rather long
for a mesophilic sewage digester and times of 12 d or so
could be used, but 20 to 30 d is required for the
digestion of some agricultural wastes. The digester
designs for agricultural wastes, particularly animal
excreta, have generally been similar to the sewage
digester: i.e., single-stage, continuous-flow, stirred
tanks. This type of digester starts with some form of
feedstock holding tank, mainly to provide some reserve
capacity in case of breakdowns and to prevent pump inputs
running dry. The feed is pumped into the digester tank
which can be made of steel or concrete and be above or
below ground. To ensure a good mixing of the input with
the tank contents, especially the active bacteria, the
feed is usually taken in towards the centre of the tank.
Most digesters consist of a cylindrical tank of low
aspect ratio, although a few are rectangular and some
municipal digesters egg-shaped. Mixing of the digester
contents is necessary only to ensure biological homo-
geneity and to give uniform heating, and is not required
for substrate mixing, as in the case of aerobic fermenters.
Mechanical mixing of various kinds is used, but the tendency now is to mix by recirculating some of the reactor
gas produced, either through free-rising spargers or
through draught-tubes. The sludge is heated either by
circulating it through an external water-filled heat-
exchanger or by circulating warm water through internal
coil, flat radiator or double-skin, draught-tube, heat
exchangers. In some big digesters the output is pumped
from the tank, but most rely on a gravity overflow
through a weir or stand-pipe system which allow pressuri-
zation of the gas in the digester head-space to a few
inches water gauge. The digester may have an integral
top or it may have a floating top which acts as a gas
holder. If it has a fixed top, gas is collected in a
separate water-sealed gas holder. In either case only a
fraction of a day's production is held, the gas being
used continuously. The sludge output passes to a tank
where it is held until it is disposed of in some way.

In the municipal sewage works in particular, this holding time serves two purposes. Firstly, the sludge settles and so a thicker sludge of smaller volume for disposal can be drawn off from the bottom of the tanks and the supernatant liquid can be returned to the sewage inflow for further purification. Secondly, slow bacterial activity continues which improves the quality of the sludge so far as pollution control is concerned.

The main objective of the sewage works digester is to turn an objectionable, polluting, sewage sludge into a stabilized, much less polluting, digested sludge. Gas production is a secondary feature and this varies with the input solids concentration and digester efficiency. In most cases the gas is used to heat the digester. With a small plant this is done by burning the gas to heat water for the digester heat exchangers. In the larger plants the gas is used in piston or gas-turbine engines which drive electric generators, air compressors and pumps to provide power for the sewage works and the aerobic treatment plants. Cooling water from the engines is then used to heat the digesters. Some sewage works produce excess power which can be used outside the works.

With the increasing concern about conventional energy supplies and their constant increase in price, attention has changed from the control of pollution due to domestic and industrial wastes to the generation of power from these wastes or the generation of gas from crop wastes or crops grown on "energy farms". Most of these feedstocks are either thick slurries or can be made into slurries. The digesters used are of a similar design to the domestic digester, although construction details may be different.

Some factories produce high volume and high flow-rate waste waters. Although these contain little suspended matter and the dissolved material may be rapidly fermented, in the case of sugars, there is still a limiting factor, due to the growth rate of the methanogenic bacteria, which imposes a detention time of some days in order to avoid wash-out of the bacteria. With flow rates of hundreds of gallons a minute even a few days' detention time calls for an unrealistically large digester tank and, as substrate levels are low, gas production per digester volume is uneconomically small. The answer to this problem lies in returning the bacteria to the digester in the "feedback" or "contact" digester, or retaining the bacteria in the tank by growth on a fixed support (the "anaerobic filter") or in a flocculent mass which can be kept in the tank by gravity and baffles (the "sludge-blanket" digester). The contact-digester tank is similar to a single-stage digester but the overflow passes through some form of separator which removes the bacteria for recycling to the tank (this is technically difficult because of gas bubbles which cause the bacteria to float). The filter and the sludge-blanket both have an upwards waste-water flow. In all three types a high liquid flow rate (giving a detention time up to a few hours) is combined with a long bacterial detention time (some days) and a high bacterial concentration, promoting the rapid digestion of the waste in a reasonably small tank size.

Batch digestion, where a container of the substrate is allowed either to develop a methanogenic flora from bacteria already present (for instance animal excreta) or is inoculated from working digester contents, is possible. However, although laboratory-scale, batch digestion is used in testing possible substrates, full-scale batch digesters are not normally used. There can be problems in the development of the correct microbial flora and digestion may be inhibited by the rapid formation of acid [Hobson and Shaw, 1973]. High concentrations of ammonia in the feedstock may totally inhibit digestion [Wong-Chong, 1975]. Another problem is that gas production follows the normal time curve for the growth of bacteria and so varies from day to day (a small batch digestion might take 30 to 40 d to complete). To obtain a constant gas production for commercial use (and to deal with a regular flow of feedstock), a number of batch digesters all out of phase at various stages of development, need to be used. Jewel [1979] suggested the use of very large batch digesters with a feedstock of, say, vegetation plus animal excreta compacted to a solids content of 30 to 40% (TS) and fermenting over a period of 100 to 365 d. His calculations suggested that such a digester could be self-heating due to the heat produced from bacterial metabolism. In digesters continuously treating slurries and liquids, the metabolic heat is insufficient to meet the needs of the digester, although it may retard cooling.

Thus, a number of types of digester are possible, but the type of digester is indicated by the feedstock and by economic considerations, rather than the bacterial reactions. The latter, so far as a limited number of investigations show, seem to be basically the same in all digestions, although the nature of the feedstock determines which of the hydrolytic and fermentative reactions predominate. Sufficient detailed examinations of digestions have not yet been done to show whether species of bacteria are common to all digesters. However, by analogy with the rumen, some of the results obtained so far show common species do occur. Added to this there is the fact that it is fairly general practice to inoculate experimental and full-size digesters one from another, or to use municipal digester contents or digesting animal waste as inocula for factory waste or vegetable-matter digesters and so on. The two systems in which there do seem to be differences are mesophilic and thermophilic digestions where different bacteria carry out the same reactions. Again, detailed studies have not been made, but the limit of mesophilic digestion is about 45°C and thermophilic digestion takes place over the temperature range about 50° to 65°C, with an optimim about 60°C; although there seem to be some differing data on this latter point. A distinct flora needs to be developed for the thermophilic range, but as this can be developed from a mesophilic digestion or from rumen contents, for instance by slow adaptation, it is not completely certain whether adaptation or selection processes take place. Unless otherwise mentioned the results quoted later are taken from work on mesophilic digestions.

3. The Overall Metabolic Picture

While digestion is a means of reducing the pollution caused by various wastes, it is being regarded more and more as a means of converting waste materials into an energy source (the gas) and a digested sludge which has fertilizer value or value as an animal feed additive (largely from the microbial protein) or some other use. Even in the case of the so-called "energy farm" projects better overall economy may be obtained if the crop can provide an animal or human foodstuff or, say, timber, and the crop residues used for digestion rather than the whole crop be grown solely for digestion. The digester feedstocks can be considered to have carbohydrate, lipid, protein, non-protein-nitrogenous compounds, salts and the undigestible materials mentioned before. The breakdown of carbohydrates, and to some extent lipids, are the energy-producing reactions of the first stages of the process. While deaminative reactions occur there is no evidence so far of Stickland-type reactions which could produce growth energy for bacteria. Nitrogen compounds are interconverted to provide cell nitrogen for new bacterial growth. Ions from salts are incorporated into bacterial cells in the provided forms or in reduced forms. The fermentation products from the carbohydrates then form the substrates for a second set of bacteria which inter-convert various organic acids and finally produce methane. In each step of the process the bacteria are interdependent and the reactions balanced. Not only is pH or other control unnecessary in a digester, but the value of the products formed precludes on economic grounds the sophisticated control mechanisms of most industrial fermenters. In addition the sites of digesters, on farms or similar places, means that attention to the fermentation process must be minimal.

Obviously, the most efficient use of substrates can be obtained if sources of energy, carbon and nitrogen are present in balanced amounts and none is in large excess. In most industrial fermentations a pure culture is used and, even if a constituent of the medium is only partially defined, the medium can be constituted to make maximum use of all the constituents. In digestion the substrate generally has to be taken as it comes. Some work has been done on mixing feedstocks to provide a better balance of carbon and nitrogen [Hills, 1979]. Excreta have a high nitrogen content and can usefully be mixed with vegetable wastes, for instance potatoes [Summers and Bousfield, 1976] or domestic refuse [Ross, 1965], containing more carbohydrate. However, carbon:nitrogen ratios cannot always be calculated purely from analysis, since the important factor is the carbohydrate and nitrogen available to the bacteria and these are usually unknown fractions of the totals determined by chemical analysis.

Generally only a fraction, usually about a half, of the apparent substrate, that is the organic matter of the feedstock, is available for digestion. The output contains carbohydrates and nitrogenous compounds which are not degraded because of their structure or because of protection by vegetable lignin, silica, waxes and other polymeric materials. In addition there are substances potentially available which are not utilized because of the imbalances mentioned before: for instance high levels of ammonia in animal excreta.

In the next sections the bacteria concerned in these reactions are discussed, but here the emphasis is placed on the ecological aspects of reactions of the digester flora rather than the detailed biochemistry. This latter aspect and more detailed descriptions of digesters and feedstocks used or laboratory-tested, are dealt with in various other reviews [Kirsch and Sykes, 1971; Hobson et al., 1974, 1980; Hobson, 1979; Bryant, 1979]. However, there have been few detailed studies of the overall microbiology of digestion. More knowledge of the reactions, and the bacteria involved in the reactions, leading to methane formation has been acquired over the last few years, but little is known about the hydrolytic and fermentative bacteria or how bacterial populations vary with different feedstocks or the differences between mesophilic and thermophilic digester populations.

4. The "Passenger" Bacteria

The microorganisms of the digester population are all bacteria and there do not, in general, appear to be any ciliate protozoa as there are in the rumen. In digesters being fed with faecal wastes, there is a large population of facultatively anaerobic and aerobic bacteria, some of which may be defined as "passengers". That is, these are organisms, including the flagellate protozoa which are sometimes seen, which cannot compete successfully in the digester environment, but which are continuously added to the digester contents with the feedstock. It is likely that ciliate protozoa are introduced in a similar fashion into digesters treating abattoir waste which would include rumen contents, and the stored, ensiled, vegetation feedstocks of proposed energy farms would contain many types of bacteria. The influence of the digester feed on such organisms can be seen from tests on a digester started from digesting municipal sludge and gradually being acclimatized to a feed of piggery waste [Hobson and Shaw, 1974]. *Escherichia coli* is a predominant facultatively anaerobic bacterium of the human intestines and is found in large numbers in human faeces. Streptococci are dominant in pig intestines and faeces. The municipal digester sludge contained *Escherichia coli*, but these were supplanted by streptococci as the digester adapted to the pig waste.

The number of facultatively anaerobic bacteria in pig excreta is high: 2.4×10^8 organisms ml^{-1} were found in piggery slurry of about 4% TS. These numbers were higher than those in the digester (5.7×10^6 organisms ml^{-1}). This bears out the suggestion that many of these bacteria are "passengers" and tend to die off in the digester although their apparent viability is kept high by the numbers entering the digester. In pig-waste digesters the number of facultatively anaerobic bacteria was similar to the anaerobic bacteria [Hobson and Shaw, 1971, 1974]. On the other hand the facultative and aerobic bacteria in municipal sewage digestions were present between 1 and 10% of the level of the anaerobic bacteria [Torien et al., 1967; Mah and Susman, 1968]. By analogy with the rumen it may be suggested that this is due to the difference in types of facultative bacteria previously noted. In the rumen growth of *Escherichia coli* is suppressed while streptococci form a part, although usually minor, of the true rumen flora, with numbers around 1×10^6 organisms ml^{-1}.

While constant addition of bacteria in the feed may, in one sense, make these facultative bacteria part of the digester flora, other experiments [Summers, 1978] have shown that it is difficult, if not impossible, to change the established hydrolytic, anaerobic flora of the digester. A rumen bacterium of high cellulolytic activity was inoculated into a pig-waste digester operating with a detention time of 10 d. One week after inoculation, the cellulolytic activity of the digester contents had increased by 14%, but after 3 weeks it had returned to the original level. Repeated tests gave the same result. The result is much the same as that obtained in similar experiments with rumen bacteria: an initial growth and then washout. However, the faster turnover of the rumen system produced more rapid washout [Hobson and Mann, 1961]. While there may be bacteria in these habitats which are not normal members of the flora and whose type can be changed by inoculation, the flora which has built up to carry out the complex reactions of substrate metabolisms appears to be very resistant to change. Presumably, natural selection has produced the best symbiotic association and the bacteria added have been rejected during the selection process.

Although the facultative bacteria seem to have no hydrolytic activities of use in degrading digester substrates, they may metabolize sugars produced by the hydrolytic bacteria and it is possible that some have a role in the scavenging of oxygen which gets into the system and in the reduction of the redox potential (Eh) to a level which is suitable for the growth of the anaerobic bacteria, in particular the methane bacteria which require an Eh of the order of -250 to -300 mV. Once the Eh has been reduced, the formation of sulphides and other reduced compounds helps to maintain low values. No mention was made earlier of special precautions taken to preclude air or reduce concentration of oxygen in the feedstock media. In fact such precautions are not needed. The digester has to be gas tight to retain the methane, but this is achieved, in floating roof digesters particularly, only with water or sludge seals, and air can get into feedstock tanks and channels. Most gas from large digesters contains a small amount of nitrogen (1 to 5% v/v) and this is almost certainly derived from air getting into the digester as there is no evidence for nitrogen production by the bacteria of digesters. Farm waste digester populations, and no doubt others, can metabolize large amounts of oxygen. For instance, an air leak into the sludge circulation pump line of an experimental pig-waste digester resulted in a composition of the gas in the head space of (v/v):71.7% nitrogen, 5.2% methane, 23.1% carbon dioxide before the leak was detected and closed [Hobson et al., 1974]. It has been found that the contents of these digesters can be pumped out, and stored in open tanks and remain anaerobic below the surface of the liquid. On returning the sludge to the digester and reheating, gas production quickly restarted. On the other hand, the anaerobic digester bacteria can be grown in pure culture only in chemically-reduced media with oxygen rigorously excluded from the gas phase and from the liquids during media preparation. Gases used to fill media tubes and bottles have to be absolutely free of oxygen. In the case of digesters, flushing the head space with an inert gas during the initial digestion stages is suggested only as a safety precaution, not as a necessity to exclude oxygen from the bacteria.

5. The Breakdown of Polymeric Carbohydrates

As previously explained, the fermentation of carbohydrates is one of the main pathways for the production of volatile acids, hydrogen and carbon dioxide, all required for methanogenesis. The amount of carbohydrate in different feedstocks varies. In most feedstocks, carbohydrates are present mainly in the form of polymers, principally as designated cellulose. It is termed "designated cellulose" since, as in the case of animal feeds, this is a material largely defined by the method of analysis rather than by chemical constitution. Cellulose may range from the prepared cellulose of "soluble' tissue papers in domestic waste to highly-lignified fibres of old vegetable matter in agricultural wastes.

The feedstock for municipal digesters usually contains a low percentage of cellulose (about 4% (w/v) was given by Pohland, 1962), consisting of residues of toilet and similar papers and the remains of cooked vegetables from human faeces. This cellulose is relatively easily degraded. The lack of cellulose in digesting municipal wastes is reflected generally by low counts of cellulolytic bacteria. Hungate [1950] found 0.8 to 2.0 x 10^3 cellulolytic bacteria organisms ml^{-1} of digesting sewage sludge and Hobson and Shaw [1974] found 4.0 x 10^3 organisms ml^{-1}. Maki [1954] found higher counts (1.6 x 10^4 to 9.7 x 10^5 organisms ml^{-1}) which is probably a reflection of differences in the composition of the feed sludges. When the sewage digester sludge of Hobson and Shaw [1974] had been adapted to piggery waste containing about 16% (w/v) cellulose the number of cellulolytic bacteria in the digester increased to 4.0 x 10^5 organisms ml^{-1}. As in the rumen system it is unlikely that cellulolysis in anaerobic digesters is carried out by only one bacterium, although *Clostridium thermocellum* is the only cellulolytic bacterium so far identified in thermophilic digesters [Ng et al., 1977]. Hobson and Shaw [1971, 1974] found 11 types of cellulolytic bacteria in their mesophilic pig-waste digester. All but one was Gram-negative although the types were not thoroughly characterized and given species names. Hungate [1950] found more than one type in sewage digesters and Maki [1954] isolated 10 different bacteria strains from a sewage digester in long-term experiments.

A characteristic of these cellulolytic bacteria is that they have different rates of attack on one particular cellulose, and this is also a characteristic of cellulolytic bacteria from the rumen. With bacteria from the latter source van Gylswyck and Labuschagne [1971] found that some strains grew as fast on cellulose as on cellobiose, whereas other strains grew faster on the cellobiose and so in these cases the rate of cellulolysis would control the growth rate when cellulose was the substrate. Similarly *C. thermocellum* from a thermophilic sewage digester grew some 3.5 times as fast on cellobiose as cellulose [Ng et al., 1977].

The rates of cellulose fermentation in pure cultures have been measured using purified materials, such as powdered filter paper, as the substrates. The substrates in sewage sludges may be very similar to these, but the lignified cellulosic substrates of animal wastes are more difficult to degrade and also contain areas of more easily (for instance the ends of fractured fibres) and less easily (for instance nodel material) degradable material as well as areas composed of different polysaccharides, such as cellulose, hemicellulose and pectin. While the paper cellulose in sewage sludges may be more or less completely fermented, the cellulose and hemicellulose of feed residues in the animal wastes and plant residues, now being used as digester feedstocks, are only partly degraded. In piggery waste only some 41% of the cellulose and 48% of the hemicellulose was degraded [Summers and Bousfield, 1980] in mesophilic digesters with a 10 d detention time. Only 35% of the straw mixed with pig waste was degraded with a 20 d detention time [Hobson, 1979]. Pfeffer [1979] extrapolated the results from thermophilic digestion of straw and calculated that the degradation at an infinite detention time could be about 50%. Unpublished calculations from the results (such as those quoted by Summers and Bousfield, 1980) of piggery waste digestions in the author's laboratories suggest that there are at least two bacterial degradations occurring in the solids in one of which the bacteria have a maximum specific growth rate of about 0.34 d^{-1} and in the other a rate of about 0.13 d^{-1}. Thus, an 8 d detention time is the minimum required for some degradation of the more recalcitrant solids and 10 to 15 d detention times are recommended for practical digestion of pig wastes. Electron microscope studies with rumen bacteria have shown that anaerobic fibre degradation involves bacterial attachment to the fibres and that bacteria preferentially degrade particular portions of the fibre structure. Bacteria can also be seen attached to fibres in animal waste digesters. A consortium of bacteria is probably involved in complete breakdown of the fibres in animal waste digesters. Hobson and Shaw [1974] found hemicellulolytic bacteria in piggery-waste digesters in numbers similar to those of the cellulolytic bacteria. Unlike the cellulolytic bacteria the former appeared to belong mainly to one species identified with the rumen *Bacteroides ruminicola*. A Gram-negative rod was also isolated in smaller numbers.

There is much more work to be done on the breakdown of fibres in anaerobic digesters. However, the few experiments completed suggest that the rates of cellulose breakdown by isolated bacteria do bear some relationship to rates observed in digesters, and one experiment suggests how symbiotic relationships in the mixed digester populations may increase rates. Maki [1954] found the maximum rate of cellulose (ground filter paper) hydrolysis by one isolate from a sewage digester in pure culture to be 260 mg cellulose l^{-1} d^{-1}. When a non-cellulolytic *Clostridium* sp. from the digester was grown with the cellulolytic bacterium the rate of cellulose degradation increased to 660 mg cellulose l^{-1} d^{-1}. This rate approached that previously reported by Buswell [1936] for cellulose digestion by whole digester sludge *in vitro* of about 1000 mg cellulose l^{-1} d^{-1}. The rate of cellulose fermentation in piggery waste digesters can be calculated from the results quoted by Summers and Bousfield [1980], when the feed contained 6% TS waste, and amounted to about 380 mg cellulose l^{-1} d^{-1}. This figure is of the same order as those quoted above, although as previously pointed out the cellulose may be more resistant to degradation.

The reason for the synergistic action of the clostridium could not be determined in the experiments mentioned above. If the hydrolysis of cellulose was not the rate limiting step for the cellulolytic bacterium, it is possible that removal of excess cellobiose was important, since cellobiose has been shown to inhibit cellulolytic activity [Hungate, 1966]. However, Maki [1954] found only glucose in the free sugars remaining in a pure culture when excess cellulose was present and hydrolysis was allowed to continue after bacterial growth had stopped. On the other hand the medium used contained complex nitrogen sources and breakdown of amino acids by the *Clostridium* sp. could have produced ammonia and branched-chain volatile fatty acids which might have stimulated growth of the cellulolytic bacterium. The role of volatile fatty acids as growth factors for the ammonia-utilizing rumen cellulolytic bacteria is now well documented. Ammonia is probably a major source of nitrogen for fermentative digester bacteria [Hobson and Shaw, 1974] but the effects of volatile fatty acids have not been extensively studied. However, tests such as those of Maki [1954] showed that digester fluid contained growth factors for digester cellulolytic bacteria and digester fluid is generally incorporated in media for isolation of the digester bacteria. The digester fluid could contain growth-stimulating volatile fatty acids as does rumen fluid or it might contain metabolites produced by some bacteria and stimulatory to others. For instance, although B_{12} is known to be formed in digesters the role of vitamin B_5 in bacterial interactions is still open to investigation [see Slater, this volume].

The ammonia-utilizing digester bacteria probably use carbon dioxide as a partial source of cell carbon as do some rumen bacteria, but the hydrogen-utilizing methanogens (or at least those that have been tested) can use acetic acid as a major source of cell carbon [Bryant, 1979].

Starch is not a substrate likely to be found in most digesters as starch in foods is rapidly utilized by all animals and man, and faecal residues are very small if any. However, some factory wastes contain starch and if crops are used as digester feedstock it is possible that these may contain starch. Waste potatoes were, for instance, suggested as a possible seasonal supplement to farm-waste digesters and experiments [Summers and Bousfield, 1976 and unpublished observations] showed that these were readily fermented when added to a pig-waste

digester without any adaptation period being required.
This is in accord with cultural studies as the pig-waste
digesters were found to contain large numbers of starch-
fermenting bacteria of different genera. *Bacteroides*
spp., *Clostridium butyricum* and aerobic Gram-negative
coccobacilli were amongst the predominant group [Hobson
and Shaw, 1971, 1974]. Torien [1967] also isolated
amylolytic clostridia from sewage sludge digesters, as
well as other amylolytic bacteria. Analysis of the pig-
waste showed the absence of starch and starch has not
been reported in analyses of sewage sludges, but amylolytic activity seems to be a common bacterial property
and no selection of an amylolytic microflora is required
in order to deal with a sudden input of starch to a
digester.

6. Degradation and Utilization of Nitrogen Compounds

Digester feedstocks contain protein and non-protein
nitrogen compounds. Proteolytic activity has been demonstrated in digesters using a number of feedstocks but
input and output analyses, for instance for pig-waste
digesters [Hobson and Shaw, 1973; Summers and Bousfield,
1980], shows that little if any overall change in the
amounts of these compounds takes place. Breakdown of
protein and deamination of amino acids must be balanced
by bacterial cell synthesis. The fermentative bacteria
isolated by Hobson and Shaw [1971, 1974] could utilize
both amino acids and ammonia, whereas the methanogenic
bacteria use only ammonia.

Unlike the rumen, where proteolytic bacteria are
principally Gram-negative, the most numerous proteolytic
bacteria isolated from sewage and animal-waste digesters
have been *Clostridium* spp. [McCarty et al., 1962;
Siebert and Torien, 1969; Hobson and Shaw, 1971, 1974].
Why this should be is not known. On the other hand
Hobson and Shaw [1974] suggested that deaminative activity
in digesters could be associated with the Gram-negative
Bacteroides ruminicola, as in the rumen.

In addition to the proteolytic clostridia other strains
of anaerobic bacteria without ability to hydrolyse polysaccharides, but with a general fermentative activity for
sugars, have been isolated. These form a heterogeneous
population, like the facultatives, probably obtaining
energy from excess sugars produced by the hydrolytic
bacteria.

7. Some Miscellaneous Reactions

Some of the bacteria isolated by Hobson and Shaw
[1974] produced lactic and succinic acids, but analyses
showed that these acids were not detectable in the
digester fluid [Hobson and Shaw, 1973]. Hobson et al.
[1974] isolated lactate-fermenting streptococci (similar

to *Streptococcus lactis*), *Bacteroides* spp., *Clostridium* spp. and one curved rod tentatively identified as *Desulfovibrio* sp. although their properties did not correspond exactly with known species. The total number of bacteria was about 3.0×10^7 organisms ml^{-1} in a pig-waste digester. All the bacteria produced acetic and propionic acids from lactate and fermented sugars.

Metal sulphides are precipitated in digester fluids, which is one reason why copper added to pig feeds, and other heavy metals in digester feedstocks are not necessarily toxic to digesters. Hydrogen sulphide is also a constituent of digester gas, although usually at less than 1% (v/v). Some of the carbohydrate-fermenting anaerobes produce hydrogen sulphide from sulphur amino acids, but the reduction of sulphate in digester feeds also takes place. *Desulfovibrio* spp. can be found, as mentioned above, in digester contents. For instance, Torien et al. [1968] found 3.0 to 5.0×10^4 sulphate-reducing bacteria ml^{-1} of fluid from a sewage digester, of which many were identified as *D. desulphuricans*. Additions of mine water containing high levels of sulphate to the digester increased their numbers from 6.6×10^3 to 9.5×10^7 organisms ml^{-1}. Thermophilic strains of *Desulfotomaculum* and *Desulfovibrio* have also been isolated from thermophilic digesters [J.G. Zeikus, pers. comm.].

Production of large amounts of free sulphide ions is undesirable in digesters since they are generally toxic to digestion processes [McCarty, 1964]. The sulphate-reducing bacteria can also grow on conjunction with hydrogen-utilizing methanogenic bacteria [Bryant et al., 1977] when sulphate is present at limiting concentration. Competition for hydrogen by sulphate-reducing bacteria in the presence of excess sulphate has been suggested as the basis for inhibition of digestion when the feedstock contains large amounts of sulphate [Winfrey and Zeikus, 1977]. On the other hand McCarty [1964] showed that added sulphide inhibited the digestion processes. Removal of sulphide ions as hydrogen sulphide in a vigorously-gassing digester or precipitation of sulphide by the addition of heavy metal salts overcomes the inhibition. The effect of iron salts, which sometimes increase gas production when they are present, may be due to sulphide precipitation.

Normally lignin is not attacked in digesters, a situation which is similar to other anaerobic environments such as the rumen. However, there is often some slight loss of lignin although how much of this is due to deficiencies in the analytical processes is uncertain. Lignin in plant material is heterogeneous and probably small molecular weight fragments can be split off during the degradation of the vegetable matter. Aromatic monomers of the types making up the lignin polymer can be degraded to give methane by enrichments of mixed digester bacteria, but the complete consortium of bacteria involved in these processes has not yet been identified [Healy and Young, 1979].

8. Lipid Metabolism: Production of Methane

Lipid forms a varying proportion of the feedstocks of anaerobic digesters. It is generally in large amounts in sewage sludges (about 30% (w/v) of the solids) but in small amounts (5 to 15% w/v) in animal-waste digesters. Analyses of input and output show that this lipid is lost during digestion, although a small amount, about 4 or 5% (w/w) is always left as lipids of the viable bacteria present in the digester. Early studies showed that the glycerides were hydrolysed and unsaturated long-chain fatty acids were hydrogenated [Heukelekian and Mueller, 1958]. Torien [1967] isolated vibrios from sewage digesters which could hydrolyse sunflower oil, and Hobson et al. [1974] isolated *Anaerovibrio lipolytica* the predominant rumen lipolytic bacterium and found levels of 1.0×10^4 to 1.0×10^5 organisms ml^{-1} of pig-waste digester fluid. These bacteria all hydrolysed long-chain and short-chain glycerides. Other bacteria, including *Bacillus* spp., with only esterase activity are also present in digesters. A number of bacteria in the pig-waste digesters could ferment the glycerol arising from lipid hydrolysis.

Experiments with whole digesting sludges [Heukelekian and Mueller, 1958; Chynoweth and Mah, 1971] showed that long-chain fatty acids were degraded to acetic acid and hydrogen. This is an energetically unfavourable reaction for bacterial growth and more recent investigations have shown that two bacteria are required to bring about the reaction. One mixture involves a bacterium which degrades C_4- to C_8-fatty acids producing acetate and hydrogen, or acetate, propionate and hydrogen and a hydrogen-utilizing methanogen [McInerny et al., 1979]. The degradation of propionate is more unfavourable than degradation of butyrate or higher acids and propionate accumulates in digesters in which methanogenesis is impaired. This can be a kind of autocatalytic inhibition [Hobson et al., 1974] as propionate is also inhibitory to hydrogen-utilizing methanogenic bacteria, for instance *Methanobacterium formicicum* [Hobson and Shaw, 1976]. Quite recently a two-component culture in which a new species of propionate-degrading bacterium (named *Syntrophobacter wolinii*) grew with a sulphate-reducing bacterium was obtained from a sewage-digester sludge inoculum incubated with propionate. The bacterium could not be grown with a hydrogen-utilizing methanogen and could not be separated from the sulphate-reducer [Boone and Bryant, 1980]. The two-component culture produced acetate and presumably hydrogen and carbon dioxide from propionate.

These are the first microbial communities to be

described in detail, but associations of two, or possibly three, bacteria forming methane from propionic or butyric acid have been isolated before [Hobson et al., 1974], and pure cultures producing methane from volatile fatty acids have also been claimed [Buraczewski, 1964]. It is likely that these were mixed cultures, as was found to be the case with *Methanobacillus omelianskii* which was supposed, for many years, to be a pure culture producing methane from ethanol [Bryant et al., 1967].

These mixed cultures demonstrate one of the major reactions determining the overall process of anaerobic digestion, the utilization of hydrogen formed by various fermentations and the shift in fermentation products or the effecting of unfavourable reactions brought about by the removal of hydrogen. Removal of hydrogen from a fermentation has been shown, with rumen bacteria, to change the products of fermentation of sugars from the reduced acids (e.g. butyric and lactic acids) and ethanol to the less-reduced acetic acid. Initially demonstrated with *Clostridium cellobioparus* and *Methanobacterium ruminantium* [Chung, 1972], the hydrogen-pulling effect has since been shown with mixed cultures of sugar fermenters and methanogenic or other hydrogen-utilizing bacteria. The sugar-fermenting bacteria, such as those previously described form various mixtures from the organic acids, ethanol, hydrogen and carbon dioxide. However, in a previous review [Hobson, Bousfield and Summers, 1974] it was argued, from the evidence of residual acids in digesters during the start-up period, the acids found when methanogenesis had been decreased and the gases also formed, that fermentation products other than acetate, hydrogen and carbon dioxide would be unlikely to be formed in large amounts in digesters. Although it was realized that some bacterial system degrading propionate and butyrate to methane was present, and the presence of lactate-fermenting bacteria has already been mentioned, it was thought that methane production from acids other than acetic was minimal. The exact contribution of the various hydrogen-removal reactions to acetate production is not known. The evidence, for instance that Cohen et al. [1979] found in a "two-phase" laboratory digester with a synthetic, glucose-containing substrate, that a mixture of acids and hydrogen was formed in the initial, fermentative stage when the dilution rate was too high for methanogens to grow properly and that methane was formed from these mixed fermentation products in the second stage (with a longer detention time), is inconclusive. The first stage of the "two-phase" digestion shows that the mixed fermentative bacteria behave as pure cultures when fermentation hydrogen is not utilized, and the second stage shows mixed fermentation products are used by a mixture of methanogens, but it does not show what happens in the complete digestion.

The isolation methods for co-cultures using higher acids involve enrichments and the counting of the bacteria is not possible. Nevertheless, it is possible to get at least an estimate of such bacteria using dilution cultures with the acids as sole substrates and without hydrogen in the culture atmosphere. Methane production is then detected and used as a method of determining acid utilization. Heukelekian and Heinemann [1939], for instance, using a most probable number method found 6.0×10^2 to 2.5×10^7 butyrate utilizers ml^{-1} in digesting sewage sludge. Hobson and Shaw [1971, 1974] found 2.0×10^4 bacteria ml^{-1} producing methane from butyrate in pig-waste digesters: this was a factor of 10 to 100 less than the counts of hydrogen-utilizing methanogens. No propionate-utilizing bacteria were found using the same techniques, but it seems possible, in view of later experiments (where 8 to 12 weeks was needed for methane production) that the 4 week incubation used in the former experiments was not long enough to enrich for the slow-growing propionate bacteria. However, Boone and Bryant [1980] obtained growth in their mixed culture in 10 to 20 d, although growth of the mixed culture without sulphate but with a hydrogen-utilizing methanogen was much slower. Of course, different bacteria may be concerned in pig- and human-waste digesters.

Whatever the relative contributions of the various possible reactions involving mixed populations of bacteria, there is no doubt that the direct precursors of methane in digestions are acetate and hydrogen plus carbon dioxide. Experiments with labelled acetate showed that 73% of the methane in sewage digestion came from acetate [Smith and Mah, 1966] and fermentation balance calculations show that, however the acetate is formed, this is close to the expected proportion.

A number of hydrogen-utilizing methanogens are now known in pure culture and hydrogen seemed, until recently, to be a universal substrate for methanogens. The exception is a thermophilic strain of *Methanosarcina* isolated from a laboratory digester [Zinder and Mah, 1979] which utilizes acetate, but other strains of *Methanosarcina* also utilize hydrogen plus carbon dioxide.

The first pure culture of a methanogenic bacterium isolated from a sewage digester, was *Methanobacterium formicicum* [Schnellen, 1947]. This organism, as the name implies, utilizes formic acid, as well as the equivalent hydrogen plus carbon dioxide and seems to be a common constituent of digester floras. Mylroie and Hungate [1954] isolated it from sewage digesters and it was the only hydrogen-utilizing methanogen isolated from pig-waste digesters, where it occurred at levels of about 1.0×10^5 organisms ml^{-1} [Hobson and Shaw, 1971, 1974]. The numbers of hydrogen-utilizing bacteria in the pig-waste digesters were 1.0×10^4 to 1.0×10^6 organisms ml^{-1} which is rather less than the numbers reported by

Mylroie and Hungate [1954] in sewage digesters: 1.0×10^5 to 1.0×10^8 organisms ml^{-1}. Smith [1966] found 1.0×10^7 hydrogen utilizers ml^{-1} in sewage digesters. Other hydrogen-utilizers, including *Methanobacterium runimantium* which was first isolated from the rumen, have been isolated from sewage digesters.

The facts presented here show that bacteria capable of converting hydrogen plus carbon dioxide to methane are found in mesophilic digesters and in numbers comparable with those of the hydrolytic and fermentative bacteria.

As digestion was established by slow addition of pig waste to a (continuous culture) stirred-tank digester initially filled with water, hydrogen and carbon dioxide were the first gases formed and methane was only produced later. The rate of methane production gradually increased and was paralleled, over some weeks, by increasing numbers of hydrogen-utilizing methanogens [Hobson and Shaw, 1971]. Similarly, fermentation initially produced high levels of acids (acetic, propionic and butyric) which gradually decreased as methanogenesis started, leaving low concentrations of acetic acid only once the digestion was established. When a commercial digester is being started care must be taken to see that substrate levels are initially kept low or the growth of fermentative bacteria may be such that the levels of acid in the digester become too high, preventing methanogen development. Both the methanogens and the fermentative bacteria are inhibited, and the digestion goes sour and may stop completely. The growth of methanogenic bacteria occurs over a restricted pH range close to pH 7.0 and the cellulolytic and similar bacteria cannot grow below pH 6.5 or slightly lower. However, when all reactions are balanced a digester is self-buffering at approximately pH 7.2.

In thermophilic digestions only one bacterium using hydrogen plus carbon dioxide has so far been described: namely *M. thermoautotrophicum* [Zeikus and Wolfe, 1972]. This was isolated from a mesophilic sewage digester sample by enrichment at a culture temperature of 65 to 70°C and so its numbers in thermophilic sewage digesters are not known.

The position with regard to acetate degradation is less clear in terms of the types of bacteria involved. As mentioned before, it is possible to determine the numbers of particular kinds of bacteria in digesters without isolating the bacteria, and this has been done for those involved in the acetic acid to methane reaction. Early on, Heukelekian and Heinemann [1939] counted 2.5×10^3 to 2.5×10^7 acetate utilizers ml^{-1} of digesting sewage sludge and Smith [1966] reported 1.0×10^5 to 1.0×10^6 bacteria ml^{-1}. *Methanosarcina barkerii*, originally described by Schnellen [1947], is the only mesophilic, acetate-utilizing bacterium so far obtained in pure culture. Strains have been isolated from a number of sewage digestions by various workers, and a thermophilic sarcina has just been mentioned.

However, the sarcina is probably not the only acetate-utilizing methanogen. It has not been observed by the author and his colleagues in cultures from pig-waste digesters, where Gram-negative rods seem to be the acetate degraders, although none have so far been obtained in unequivocally pure culture [Hobson et al., 1974]. Cohen et al. [1979] showed photomicrographs of the bacteria in the methanogenic stage of their mesophilic two-phase culture where volatile fatty acids were being metabolized and these were predominantly rods of various sizes. Van den Berg et al. [1976] also found rods in their enrichment cultures of acetate methanogens developed from a pear-waste digester.

There has been controversy over the rate-limiting reactions in anaerobic digesters. As previously described, fibre breakdown is almost certainly a primary rate-limiting reaction for full degradation of farm wastes and feedstocks with similar fibrous constituents. But methanogenesis may be limiting in digesters of easily fermentable substrates and can be a secondary rate-limiting step in the digestion of animal wastes. The minimum doubling times of fermentative, mesophilic, anaerobic rumen bacteria growing on mono- or disaccharides are greater than many aerobes or facultative anaerobes, but are of the order of 2 to 3 h [Hobson, 1965; Hobson and Summers, 1967]. The digester bacteria may be similar. The minimum doubling times of the hydrogen-utilizing mesophilic methanogens seem longer generally.

There is general agreement that the maximum doubling times of the acetic acid-dependent methanogens are slower still with some disagreement on the actual times, although all are of the same order. Hobson and McDonald [1980] showed that a continuous culture of the acetic acid methanogen(s) behaved like the acid-degrading step in pig waste digesters operated at different retention times, but also at 35°C, with a maximum specific growth rate about 0.4 d^{-1}. This was in agreement with the detention time of 100 h used by Cohen et al. [1979] for the methanogenic stage of their two-phase laboratory digester and for the specific growth rate of 0.49 d^{-1} calculated by Ghosh and Klass [1978] for the washout dilution rate of the methanogenic stage of a two-phase laboratory digester. However, the saturation constant (k_s) for acetic acid degradation is high: Hobson and McDonald [1980] calculated a value of k_s = 800 to 900 mg acetic acid ml^{-1}. Thus in order to keep digestion going at rates near to these maxima, high concentrations of acid must be supplied, either artifically as in the two-phase digesters or in the feedstocks as the pig-waste digesters. In the latter case feedstock acid concentrations were high because of fermentations in the waste-storage tanks of the piggery. It was pointed out that solids degradation to acids in the pig-waste digesters and so supply of fermentation acids was falling at detention times around 7 d, but some gas production was continuing at 3 d detention time.

Methanosarcina barkerii has also been found to be a slow-growing organism, but Mah *et al.* [1978] isolated a strain of *Methanosarcina* which has a doubling time on acetate of 24 to 72 h depending on the culture conditions. Even so, the former time is much longer than the doubling times of the fermentative bacteria or the hydrogen utilizers. Thus the main conclusion stands, namely that acetate utilization is a rate-limiting step in mesophilic digestion and, although the reactions may be rather faster, the same conclusion most likely applies for the rate-limiting step in thermophilic digestions.

The propionate-degrading bacteria may be slower growing than the acetate-utilizers, but the contribution of these to the overall digestion is unknown. It must, however, be less quantitatively than direct, fermentative acetate formation and methanogenesis.

A further reaction, the significance of which has not been quantitatively assessed, is the conversion of hydrogen plus carbon dioxide to acetic acid. Chynoweth and Mah [1971] found evidence for such a reaction in some early tracer studies with digesting sludge. A bacterium carrying out this reaction has been isolated from digesters in numbers of less than 1.0×10^4 to 3.5×10^5 organisms ml^{-1}. This organism is *Clostridium aceticum* [Ohwaki and Hungate, 1977]. The reaction, if it occurs to any extent in digesters, ought to increase the proportion of methane generated from acetate, but should not alter the overall level of methane production.

9. General Observations

The general difficulties in isolating and growing the digester bacteria, particularly for methanogens, has meant that two or three component cultures for modelling the digester reactions have been difficult to set up. Weimer and Zeikus [1977] set up mixed cultures (small batch) of *Clostridium thermocellum* and *Methanobacterium thermoautotrophicum*. Cellobiose was fermented faster than cellulose by the clostridium alone or in mixed culture (cf. previous section on cellulose degradation). In the mixed culture methane was produced but only from hydrogen produced by the clostridium and acetate was produced but not utilized, although it has been suggested that bacteria such as *M. thermoautotrophicum* could utilize acetate [Zeikus *et al.*, 1975]. Rose and Pirt [1980] have reported a consortium of two mycoplasmas from an enrichment of digesting sewage sludge, one of which fermented glucose to volatile fatty acids, hydrogen and carbon dioxide and the other converted the hydrogen plus carbon dioxide to methane.

Winter and Wolfe [1979] found that *Methanosarcina barkerii* in mixed culture with *Acetobacterium woodii* formed methane. The *Acetobacterium* sp. fermented fructose to acetate which was used by the sarcina, so the carbohydrate was converted to methane.

In the introductory sections mention was made of the methods for dealing with high flow-rate wastewaters without washing-out the methanogens by using the anaerobic filter or the sludge-blanket digester. In the anaerobic filter the bacteria attach to a solid matrix. Mah *et al.* [1978] showed scanning electron micrographs of *Methanosarcina* sp. in pure culture growing in a liquid medium with calcium acetate. The sarcina attached to calcium carbonate formed granules. This is not the same, of course, as attachment to the inert matrix of the anaerobic filter or the fluidized-bed digester where the bacteria are attached to minute glass spheres or similar particles, but it does suggest that the presence of surfaces could be important in general digester bacterial ecology. Many free bacteria are seen in digesters with feedstocks of high-solids sludges, but attachment of cellulolytic bacteria to fibres has already been mentioned. Some evidence for the attachment of methanogenic bacteria to fibres and other particles of animal-waste digesters has been obtained, although no evidence was obtained in cultures for adherence, or enhancement of growth, of methanogens by the inclusion of dried fibres and other particles [S. Bousfield and R. Summers, unpub. obs.].

Where surfaces are not available the bacteria seem to grow in "conglomerates" with mutual attachment and "intertwining". In the sludge-bed digester, where waste waters (as distinct from sludges) are being treated, the bacteria grow as a flocculent mass [Lettinga *et al.*, 1980] and Cohen *et al.* [1979] showed micrographs of the flocculent masses of bacteria in both the fermentative and methanogenic stages of their two-phase fermenter where the feedstocks were particle-free solutions. Ward *et al.* [1978] found enrichment cultures of *Methanosarcina* sp. transferred for two years on an acetate medium, contained several non-methanogenic bacteria. Nutritional studies showed the "satellite" bacteria could only be growing because the *Methanosarcina* sp. produced substrates and growth factors for them (see Slater, this volume). These factors might be produced by growing sarcina or by lysis of dead cells. In the author's laboratories cultures of acetate- or hydrogen-utilizing bacteria have been carried through many transfers and dilutions in media with only minerals, ammonium salts and acetate or hydrogen and yet contaminant bacteria, in small numbers and detectable by growth on transfer to rich media containing sugars, yeast extracts and other nutrients, were found associated with the methanogens.

The substrate levels in different anaerobic digesters

vary but it will have been noted that many of the bacterial counts quoted seem low, compared with the rumen counts. While culturing and counting techniques may be at fault or bacteria on fibres may not be properly counted, a possible explanation for the low counts is the very low overall bacterial growth rates due to the long turnover times of the digesters. Detention times of 20 d are well below the minimum detention times of laboratory cultures in which yields very much below the maximum observed growth yield have been recorded because of the maintenance requirements and death of slow-growing bacteria. Where a range of counts for a bacterium in a particular digester has been given, this range represents counts over a long time period. Even with the continuously loaded digester, reactions cycle with time and gas production varies somewhat from day to day [Hobson and Robertson, 1977]. This variation is caused by slight differences in feedstock composition and by the general complexity of the individual reactions. It seems likely that a complex bacterial system, such as a digester, never attains an absolute steady state comparable with laboratory pure culture, dissolved-substrate, continuous culture systems.

10. Conclusion

All the aspects of bacterial ecology in anaerobic digestions cannot be covered in detail in a short paper. However, if this chapter were longer it would be apparent that digester microbiology lacks much basic information compared with, say, that available from the rumen systems. Nevertheless, sufficient information is now known to show that a complex ecological niche, with many interacting bacteria and bacterial systems, is present in anaerobic digesters.

It is probably the case that in all natural circumstances (and a digester is in many respects natural), the anaerobes must grow in association with other anaerobes or facultative anaerobes: it is through these relationships that they have survived over the millenia in a bioshpere primarily dependent on oxygen and sunlight.

References

Boone, D.R. and Bryant, M.P. (1980). Proprionate-degrading bacterium, *Syntrophobacter wolinii* sp. nov. gen. nov., from methanogenic ecosystems. *Applied and Environmental Microbiology* **40**, 626-632.

Bousfield, S., Hobson, P.N. and Summers, R. (1974). Pilot-plant high-rate digestion of piggery and silage wastes. *Journal of Applied Bacteriology* **37**, xi.

Brade, C.E. and Noone, G.P. (1979). Anaerobic sludge digestion — need it be expensive? I. Making more use of existing resources. Institute of Water Pollution Control, University of Aston, 1979.

Bryant, M.P. (1979). Microbial methane production — theoretical aspects. *Journal of Animal Science* **48**, 193-201.

Bryant, M.P., Campbell, L.L., Reddy, C.A. and Crabhill, M.R. (1977). Growth in desulfovibrio in lactate or ethanol media low in sulphate in association with H_2-utilizing methanogenic bacteria. *Applied and Environmental Microbiology* **33**, 1162-1169.

Bryant, M.P., Wolin, E.A., Wolin, M.J. and Wolfe, R.S. (1967). *Methanobacillus omelianskii*, a symbiotic association of two species of bacteria. *Archives Microbiology* **59**, 20-31.

Buraczewski, G. (1964). Methane fermentation of sewage sludge. 1. The influence of physical and chemical factors on the development of methane bacteria and the course of fermentation. *Acta Microbiologica Polonica* **13**, 321-329.

Buswell, A.M. (1936). State of Illinois Division of the State Water Survey. *Bulletin* **32**. (Quoted by Maki, 1954).

Chung, K.T. (1972). An ecological significance of hydrogen utilization in methanogenesis. Abstracts Annual General Meeting, American Society for Microbiology p.64.

Chynoweth, D.P. and Mah, R.A. (1971). Volatile acid formation in sludge digestion. Advances in Chemistry Series, American Chemical Society **no.105**.

Cohen, A., Zoetmeyer, R.J., van Deursen, A. and van Andel, J.G. (1979). Anaerobic digestion of glucose with separated acid production and methane formation. *Water Research* **13**, 571-580.

Ghosh, S. and Klass, D. (1978). Two-phase anaerobic digestion. *Process Biochemistry* **13**(4), 15-24.

Healy, J.B. and Young, L.Y. (1979). Anaerobic biodegradation of eleven aromatic compounds to methane. *Applied and Environmental Microbiology* **38**, 84-89.

Heukelekian, H. and Heinemann, B. (1939). Studies on the methane-producing bacteria. 1. Development of a method for enumeration. *Sewage Works Journal* **11**, 426-435.

Heukelekian, H. and Mueller, P. (1958). Transformation of some lipids in anaerobic sludge digestion. *Sewage and Industrial Wastes* **30**, 1108-1120.

Hills, D.J. (1979). Effect of carbon: nitrogen ratio on anaerobic digestion of dairy manure. *Agricultural Wastes* **1**, 267-278.

Hobson, P.N. (1965). Continuous culture of some anaerobic and facultatively anaerobic rumen bacteria. *Journal of General Microbiology* **38**, 167-180.

Hobson, P.N. (1979). Production of methane from wastes and crops, In "Processes for Chemicals from some Renewable Raw Materials" pp. 1-8. The Institute of Chemical Engineers, London.

Hobson, P.N. (1979). Straw as feedstock for anaerobic digesters, In "Straw Decay and its Effect on Disposal and Utilization" (ed. E. Grossbard), pp. 217-224. John Wiley, London.

Hobson, P.N., Bousfield, S. and Summers, R. (1974). Anaerobic digestion of organic matter. *Critical Reviews in Environmental Control* **4**, 131-191.

Hobson, P.N., Bousfield, S. and Summers, R. (1980). Methane production from agricultural and domestic wastes. Applied Science Publishers, London.

Hobson, P.N. and Mann, S.O. (1961). Experiments relating to the survival of bacteria introduced into the sheep rumen. *Journal of General Microbiology* **24**, i.

Hobson, P.N., Mann, S.O. and Oxford, A.E. (1958). Some studies on the occurrence and properties of a large Gram-negative coccus from the rumen. *Journal of General Microbiology* **19**, 462-472.
Hobson, P.N. and McDonald, I. (1980). Methane production from acids in piggery-waste digesters. *Journal of Chemical Technology and Biotechnology* **30**, 405-408.
Hobson, P.N. and Robertson, A.M. (1977). Waste Treatment in Agriculture. Applied Science Publishers, London.
Hobson, P.N. and Shaw, B.G. (1971). The role of strict anaerones in the digestion of organic material, In "Microbial Aspects of Pollution" (eds. G. Sykes and F.A. Skinner) pp. 103-121. Academic Press, London.
Hobson, P.N. and Shaw, B.G. (1973). The anaerobic digestion of waste from an intensive pig unit. *Water Research* **7**, 437-449.
Hobson, P.N. and Shaw, B.G. (1974). The bacterial population of piggery waste anaerobic digesters. *Water Research* **8**, 507-516.
Hobson, P.N. and Shaw, B.G. (1976). Inhibition of methane production by *Methanobacterium formicicum*. *Water Research* **10**, 849-852.
Hobson, P.N. and Summers, R. (1967). Continuous culture of anaerobic bacteria. *Journal of General Microbiology* **47**, 53-65.
Hungate, R.E. (1950). The anaerobic mesophilic cellulolytic bacteria. *Bacteriological Reviews* **14**, 1-49.
Hungate, R.E. (1966). The Rumen and its Microbes. Academic Press, New York.
Jewell, W.J. (1979). Future trends in digester design, In "Anaerobic Digestion" (ed. D.A. Stafford, B.I. Wheatley and D.E. Hughes). Applied Science Publishers, 1980, pp.467-491.
Kirsch, E.J. and Sykes, R.M. (1971). Anaerobic digestion in biological waste treatment. *Progress in Industrial Microbiology* **9**, 155-237.
Lettinga, G., van Velsen, A.F.M., Hobma, S.W., de Zeeuw, W. and Klapwijk, A. (1980). Use of the upflow sludge blanket (USB) reactor concept for biological wastewater treatment, especially for anaerobic treatment. *Biotechnology and Bioengineering* **xxii**, 699-734.
Lysons, R.J., Alexander, T.J.L., Hobson, P.N., Mann, S.O. and Stewart, C.S. (1971). Establishment of a limited rumen microflora in gnotobiotic lambs. *Research in Veterinary Science* **2**, 486-487.
Lysons, R.J., Alexander, T.J.L., Wellstead, P.D., Hobson, P.N., Mann, S.O. and Stewart, C.S. (1976). Defined bacterial populations in the rumens of gnotobiotic lambs. *Journal of General Microbiology* **94**, 257-269.
Mah, R.A., Smith, M.R. and Baresi, L. (1978). Studies on an acetate-fermenting strain of *Methanosarcina*. *Applied and Environmental Microbiology* **35**, 1174-1184.
Mah, R.A. and Susman, C. (1968). Microbiology on anaerobic sludge fermentation. 1. Enumeration of the non-methanogenic bacteria. *Applied Microbiology* **16**, 358-361.
Maki, L.R. (1954). Experiments on the microbiology of cellulose decomposition in a municipal sewage plant. *Antonio van Leeuwenhoek* **20**, 185-200.

Mann, S.O. (1963). Some observations on the airborne dissemination of rumen bacteria. *Journal of General Microbiology* **33**, ix.

Mann, S.O. and Stewart, C.S. (1974). Establishment of a limited rumen flora in gnotobiotic lambs fed on a roughage diet. *Journal of General Microbiology* **84**, 379-382.

McCarty, P.L. (1964). Anaerobic waste treatment fundamentals, part three, toxic materials and their control. Public Works November, 91-94.

McCarty, P.L., Jeris, J.S., McKinney, R.E., Reed, K. and Vath, C.A. (1962). Microbiology of Anaerobic Digestion. Report no. R62-29, Sedgewick Laboratory, Massachusetts Institute of Science and Technology, Cambridge, U.S.A.

McInerny, M.J., Bryant, M.P. and Pfennig, N. (1979). Anaerobic bacterium that degrades fatty acids in symtrophic association with methanogens. *Archives of Microbiology* **122**, 129-135.

Mylroie, R.L. and Hungate, R.E. (1954). Experiments on the methane bacteria in sludge. *Canadian Journal of Microbiology* **1**, 55-64.

Ng, T.K., Weimer, P.J. and Zeikus, J.G. (1977). Cellulolytic and physiological properties of *Clostridium thermocellum*. *Archives of Microbiology* **114**, 1-7.

Ohwaki, K. and Hungate, R.E. (1977). Hydrogen utilization by Clostridia in sewage sludge. *Applied and Environmental Microbiology* **33**, 1270-1274.

Pfeffer, J.T. (1979). Biological conversion of biomass to methane, In "Second Quarterly Progress Report" Dynatech R/D Company, 99 Erie Street, Cambridge, Mass., U.S.A.

Pohland, F.G. (1962). General review of literature on anaerobic sludge digestion. *Engineering Extension Bulletin, Purdue University* **20**, 583-591.

Rose, C.S. and Pirt, S.J. (1980). The roles of two mycroplasmal agents in the conversion of glucose to fatty acids and methane. *Society for General Microbiology Quarterly* **8**, 40.

Ross, W.E. (1965). Dual disposal of garbage and sewage at Richmond, Indiana. *Sewage and Industrial Wastes* **26**(2), 140-148.

Schnellen, C.G.T.P. (1947). Onderzoekingen over de methaangisting. Dissertation, Technische Hoogeschool, Delft, Holland.

Siebert, M.L. and Torien, D.F. (1969). The proteolytic bacteria present in the anaerobic digestion of raw sewage sludge. *Water Research* **3**, 241-250.

Smith, P.H. (1966). The microbial ecology of sludge methanogenesis. *Developments in Industrial Microbiology* **7**, 156-161.

Smith, P.H. and Mah, R.A. (1966). Kinetics of acetate metabolism during sludge digestion. *Applied Microbiology* **14**, 368-371.

Summers, R. (1978). The microbial breakdown of straw in the rumen and the breakdown of straw and manure for methane production, In "Report on Straw Utilization Conference" pp. 84-87. Ministry of Agriculture, Fisheries and Food, London.

Summers, R. and Bousfield, S. (1976). Practical aspects of anaerobic digestion. *Process Biochemistry* **11**(5), 3-6.

Summers, R. and Bousfield, S. (1980). A detailed study of piggery-waste anaerobic digestion. *Agricultural Wastes* **2**, 61-78.

Torien, D.F. (1967). Enrichment culture studies on aerobic and facultative anaerobic bacteria found in anaerobic digesters. *Water Research* **1**, 147-155.

Torien, D., Siebert, M.L. and Hattingh, W.H.J. (1967). The bacterial nature of the acid-forming phase of anaerobic digestion. *Water Research* **1**, 497-507.

Torien, D.F., Thiel, P.G. and Hattingh, W.H.J. (1968). Enumeration isolation and identification of sulphate-reducing bacteria of anaerobic digestion. *Water Research* **2**, 505-509.

van den Berg, L., Patel, G.B., Clark, D.S. and Lentz, C.P. (1976). Factors affecting rate of methane formation from acetic acid by enriched methanogenic cultures. *Canadian Journal of Microbiology* **22**, 1312-1319.

Van Gylswyck, N.O. and Labuschagne, J.P.L. (1971). Relative efficiencies of pure cultures of different species of cellulolytic rumen bacteria in solubilizing cellulose *in vitro*. *Hournal of General Microbiology* **66**, 109-113.

Ward, D.M., Mah, R.A. and Kaplan, I.R. (1978). Methanogenesis from acetate: a nonmethanogenic bacterium from an anaerobic acetate enrichment. *Applied and Environmental Microbiology* **35**, 1185-1192.

Weimer, P.J. and Zeikus, J.G. (1977). Fermentation of cellulose and cellobiose by *Clostridium thermocellum* in the absence and presence of *Methanobacterium thermoautotrophicum*. *Applied and Environmental Microbiology* **33**, 289-297.

Winfry, M.R. and Zeikus, J.G. (1977). Effect of sulphate on carbon and electron flow during microbial methanogenesis in freshwater sediments. *Applied and Environmental Microbiology* **33**, 275-281.

Winter, J. and Wolfe, R.S. (1979). Complete degradation of carbohydrate to carbon dioxide and methane by syntrophic cultures of *Autobacterium woodii* and *Methanosarcina barkerii*. *Archives of Microbiology* **121**, 97-102.

Wong-Chong, G.M. (1975). Dry anaerobic digestion. In "Energy, Agriculture and Waste Management" (ed. W.J. Jewell), pp.361-371. Ann. Arbor Science.

Zeikus, J.G., Weimer, P.J., Nelson, D.R. and Daniels, L. (1975). Bacterial methanogenesis: acetate as a methane precursor in pure culture. *Archives of Microbiology* **104**, 129-134.

Zeikus, J.G. and Wolfe, R.S. (1972). *Methanobacterium thermoautotrophicus* sp.n. an anaerobic, autotrophic, extreme thermophil. *Journal of Bacteriology* **109**, 707-713.

Zinder, S.H. and Mah, R.A. (1979). Isolation and characterization of a thermophilic strain of *Methanosarcina* unable to use H_2-CO_2 for methanogenesis. *Applied and Environmental Microbiology* **38**, 996-1008.

Chapter 4

MIXED CULTURES IN AEROBIC WASTE TREATMENT

HUGH J. SOMERVILLE

*Shell Research Ltd,
Sittingbourne, Kent, UK*

1. Introduction

For those of man's activities which result in the widespread distribution of natural and synthetic organic compounds, degradation and recycling depend on natural processes. Some of these waste products are collected for disposal and treatment whereas others are dispensed so widely in the environment that, once distributed, their ultimate fate depends on factors beyond man's present capacity for intervention. Although the activity of microorganisms is not the only mechanism involved in recycling it makes a major contribution in many cases; a combination of physico-chemical and biological degradation [Plimmer, 1978] may be important with both dispersed and collected chemicals. For these wastes which can be collected and which are recycled by breakdown to the constituent elements a microbiological approach through a designed process is often the method of choice, mainly for reasons of economy including energy conservation. This applies particularly to purification of waste water where microbial processes are effective at the concentrations of dissolved compounds which would generally be too low for economic physico-chemical treatment. Application of microbiological waste treatment processes provides the basis for what is probably the single largest economic contribution of biotechnology, albeit one in which the technological approach has been consistently conservative for the most part.

1.1. *Industrial and Domestic Wastes*

Industrial wastes form a spectrum in terms of their amenability to biotreatment and they may differ from domestic wastes in some important factors, not all of which are relevant to waste from any particular industrial source.

Present address: CONCAVE, The Hague, The Netherlands.

Domestic wastes generally arrive at the treatment unit in a fairly predictable pattern with respect to nutrient concentration, often including a well characterized diurnal variation. Chemical wastes are often unpredictable in both qualitative and quantitative terms. Experience of the impact of sudden concentration changes in chemical waste streams has led to the incorporation of large holding or equalization tanks upstream of many biotreatment units. Such holding tanks, in which the partial mineralization of organic carbon may occur, allow the damping of any sharp changes in concentration of individual chemicals in an effluent. They may also be used to retain large-scale spills for subsequent controlled discharge.

Domestic wastes consist largely of naturally-occurring compounds and are readily degraded by microorganisms whereas chemical wastes invariably contain some more or less recalcitrant compounds.

Chemical waste-waters are more likely to include materials which can be partially oxidized to intermediates or dead-end products that are more recalcitrant than the parent compound, leading to the appearance in the effluent of compounds not present in the influent. This has been shown, in the Shell laboratories, to happen with bench-scale biotreatment systems in degradation of isopropanol and in conversion of other alcohols. This may be a general problem in the conversion of secondary or tertiary alcohols where non-specific alcohol dehydrogenase activities present in the biomass may lead to the formation of the corresponding ketones. Similarly, limited conversions may occur with aromatic chlorides [Reineke and Knackmus, 1979]. Such partial transformation can also occur in domestic sewage, for example, during the degradation of detergents [Stiff, 1979]. Incomplete nitrification can result from poor performance or overloading of domestic plants [Downing et al., 1964].

The nutrient status of domestic wastes is generally predictable and consistent compared to that of chemical wastes. In the latter, nitrogen may be completely absent or present as ammonia or combined nitrogen. The concentration of metals may also vary extremely widely. Other parameters, such as temperature, pH, dissolved oxygen and others are likely to vary widely in industrial wastes.

Although it is generally considered acceptable to "farm" the sludges (biomass) derived from biotreatment of wastes of purely domestic origin, the presence of potentially harmful substances, such as heavy metals, can limit or eliminate the application of this approach to biomass derived from the growth of microorganisms on industrial wastes.

Thus, although the biological treatment of chemical and domestic wastes both share a common objective, water purification, and common engineering technology, the microbial processes involved vary widely and any attempt to generalize on the treatment of industrial wastes must be treated with more than ordinary caution.

This contribution will concentrate on the role of mixed cultures in the aerobic treatment of mixed, collected wastes. The terms biotreatment and biotreater are taken to apply to the process and apparatus (bench or process-scale) in which organic wastes are mineralized through the action of microorganisms. Because of the relatively small amount of information on biological treatment of chemical wastes, studies on other wastes, largely municipal in origin, have been selected to illustrate specific points without the objective of a comprehensive review. More detailed reviews of the general microbiological aspects have been provided by Hughes and Stafford [1976] and Taber [1976].

2. Microbial Degradation of Collected Wastes

In the last few years there has been a revitalization in interest in the detailed microbiology of mixed cultures, microbial communities and consortia [see Slater, this volume]. Some of these studies have utilized chemostats and have indicated that catabolism of a single compound may involve several species, some of which are not able to attack the parent compound directly. Such systems, with a single carbon source and, normally, a single factor limiting growth, are simple compared to waste treatment systems where there may be many and variable limiting factors as well as variations in the spectrum of carbon compounds. Designed biotreaters may involve recycling of biomass, as in many activated sludge systems, or partial immobilization of biomass as in trickle filters and rotating discs. Thus an enormous number of different factors can act as the selection pressure. The inevitable question is that, if it takes several species to utilize single chemicals in chemostats operating under controlled conditions [Slater and Somerville, 1979], how many species will be involved in the treatment of an effluent containing many different chemicals in continually varying amounts and with very poor control over operating parameters?

For these reasons, it is not surprising that only limited success has been achieved in the characterization of microbial communities involved in the treatment of collected wastes, whether of domestic or industrial origin. Selection pressures will vary greatly both from location to location and within a single biotreater at different times. The variation in the microbial population is probably only partly reflected in the visible problems of bulking, dispersed growth and other important factors. These manifestations of variation probably reflect extremes in the operating conditions. Indeed the requirement for recycling, where this is part of the biotreatment system, and the ability to withstand the successive insults of imposed variation, are likely to be amongst the most important selection factors.

2.1. *Interactions Between Different Species*

The factors affecting interaction between species have been extensively discussed elsewhere [Jannasch, 1967; Bungay and Bungay, 1970; Veldkamp and Jannasch, 1972; Harder et al., 1977; Fredrickson, 1977; Slater, 1978; Slater and Somerville, 1979; Konings and Veldkamp, 1980; see Slater, this volume]. Clear evidence has been presented for some of the types of interaction that have been identified and could occur in biotreaters. Thus a small change in operating conditions can lead to a large change in the population. It has also been established that small changes in genotype can lead to large changes in population [Zamenhof and Eichhorn, 1967; Mason and Slater, 1979]. From these observations made with well-defined microbial populations, it is clear that description of the relationships in biotreater communities presents a formidable task to the microbial ecologist, particularly in view of the open nature of the systems.

2.2. *Microbial Species Involved in Waste Treatment*

Although the involvement of individual microbial species in degradation of specific chemicals has formed one of the bases of our present understanding of metabolic pathways, very little work has been carried out on the role of the many individual species within the complex mixture of microorganisms involved in biotreatment. Several consistent problems have emerged.

Low recoveries of viable bacteria probably arise from the presence of a large proportion of non-viable bacteria; i.e., those that are unable to reproduce but are not necessarily biochemically inactive, and/or from the failure to use conditions which allow the growth of some microorganisms as single colonies. The contribution of non-viable biocatalysts to soil processes has been discussed by Burns [1979] and they may also contribute to the degradation processes in biotreaters. Commonly, about 10% of the total microscopic count appears as a viable count.

In the experience of the author, colony counts of about 25% to 30% of the total microscopic count can be obtained by plating out, on nutrient agar, samples of activated sludge from bench-scale biotreaters utilizing chemical wastes, provided that the biomass is briefly subjected to ultrasonic disruption [Banks and Walker, 1977]. Further study [J.P. Salanitro, pers. comm.] has indicated that consistently higher recoveries (35% to 40%) can be recovered in media containing added peptides, glucose and vitamins. The main filamentous species identified are *Nocardia* (and other

Actinomycetales) and *Bacillus*. With domestic wastes Banks *et al.* [1976] have found that the viability varied in samples from a number of treatment plants. A three to four fold increase in the number of viable cells per gram of sludge appeared to be related to the calculated growth rate. These authors also observed that the bacterial population, as characterized by division into groups on the basis of biochemical tests, varied widely between different biotreaters. It is clear that very different nutritional conditions hold in the biotreater and for growth in the laboratory. In particular, the available concentration of nutrients in a biotreater should be extremely low under normal operating conditions and it is tempting to speculate that some species may not be able to reproduce, or to grow rapidly enough, to register as viable organisms under laboratory conditions. An interesting development in this area has been the recent proposal for a role of proton-motive force in viability [Konings and Veldkamp, 1980].

In a recent study aimed at the development of a constant microbial population for treating a single waste in the laboratory, Wittauer [1980] developed a key for identification of the species but found large changes in the population distribution over a period of time. Eikelboom [1975] investigated the filamentous microorganisms occurring in activated sludge from some 200 water purification plants including some that treated industrial wastes. Largely on the basis of morphological characteristics, as observed by direct microscopic examination of the sludge, 26 types were identified which were divided into seven groups. Subsequently Eikelboom [1977] proposed a key for identification based on the needs of microbiologists and plant operators. There is no doubt that this forms a useful pragmatic approach which will help in plant operation. Some of the microorganisms described in this and other work have clearly not been isolated in pure cultures. It should also be noted that some of the filamentous species associated with activated sludge can exist in non-filamentous forms, e.g., *Bacillus* and *Nocardia* species. Their morphology may be controlled by environmental conditions which have not yet been reproduced in the laboratory. In the absence of suitable isolation procedures the detailed comparison of the taxonomy of sludges from both domestic and industrial biotreaters remains to be carried out. Nonetheless, the gross morphological characteristics have much in common.

Much interest has been shown in the filamentous microorganisms present in activated sludge because of their involvement in problems of biomass separation including those which are known as "bulking". This term "bulking" can be widely interpreted to mean sludge that does not settle because the biomass tends to float or stay in suspension. It is one of the most common operational problems in activated sludge plants whether treating domestic or industrial wastes. Sezgin et al. [1978], working largely with domestic wastes, developed an explanation of floc formation and properties in which the settling properties depend on the proportion of filamentous forms. Their observations were that flocs did not hold together well in the presence of too small a proportion of filaments and that the flocs were too large when too many filaments were present leading to the extreme separation problems of dispersed bacteria and bulking respectively. Although there is a substantial amount of collective experience to support this simple and rather elegant explanation, other factors are undoubtedly involved, including surface charge, polysaccharide content, density and others, and the role of these factors is discussed below. There is, however, no clear explanation as yet of the factors controlling filamentation in mixed culture.

Of course the nature of the microbial communities that are involved in waste treatment varies enormously. For example, high-rate algal ponds utilize carbon dioxide and some breakdown products produced by bacteria, for which oxygen is provided in an interaction based on gaseous substrates. The algae simultaneously release oxygen to be used by the bacteria. Shelef and colleagues have extensively studied the application of such ponds to domestic wastes [Oron et al., 1979], although no reports have been made of application to industrial wastes. This photosynthetic system contrasts with conventional activated sludge systems, for example those treating carbonization liquors [Jones and Carrington, 1972; Catchpole and Cooper, 1972].

3. Physiological Aspects

Performance of biological waste treatment is generally measured simply as the reduction in biochemical oxygen demand (BOD) between influent and effluent. Generally, about 90% reduction is expected although more complete removal of BOD may be obtained, particularly where the treatment system has plug-flow characteristics. In the continuous-stirred-tank-reactor (CSTR) configuration, the effluent concentration is theoretically the same as that in the reactor. CSTR systems may have limitations with respect to the reduction of pollutant concentrations to very low levels. Assuming that the Michaelis–Menten relationship holds, with respect to substrate concentration, for the degradative activities of biomass in biotreaters, the concentration of a substance in the effluent must be at least the saturating concentration before the full potential capacity of the biomass for removal of the substance is expressed.

Study of phenol degradation by pure and mixed cultures of *Pseudomonas putida* [Yang and Humphrey, 1975] indicated that a single stage biotreater system could achieve phenol removal to 1 to 2 mg phenol l^{-1}. However, these authors pointed to the possibility of long lasting transients resulting from inhibition by phenol, which occurred at concentrations above 1 mM. Kaspar and Wuhrmann [1978] studied the continuous anaerobic sludge-digestion process and found that, under steady-state conditions, the acetate-cleaving activities were saturated to about 10 to 15% of the maximum rate and hydrogen removal was about 1% of maximum.

The important effect of concentration of chemicals on their degradation in natural environments has been widely recognized [Postgate, 1973]. Great difficulty has been experienced in predicting rates of disappearance of xenobiotic compounds and it may be that the nature of the particular natural situation is more important than any absolute rate measurement, except where a compound is apparently "completely" undegradable. Prolonged exposure of bacterial populations to low concentrations has been shown to lead to the selection of new strains. Hacking et al. [1978] showed that *Escherichia coli* mutants with improved scavenging ability for 1,2-propanediol could be selected by continuous cultivation on media containing the compound at concentrations which were less than 0.5 mM. The marked effect of concentration on the rate of biodegradation of some xenobiotic compounds has been established in batch experiments. For example Boethling and Alexander [1979] found little conversion of 2,4-dichlorophenoxyacetate at about 0.1 to 0.2 mM, although 60% conversion to carbon dioxide was observed at higher concentrations. Solubility and the degree of substitution may also effect degradation rates as illustrated by the work of Clark et al. [1979] who isolated, by batch enrichment, mixed cultures capable of degrading several polychlorinated biphenyls. The isomers that were more water soluble and had fewer chlorine substituents were degraded more rapidly. A useful correlation has been made recently between chemical hydrolysis and biodegradation rates for some pesticides [C. Steen, pers. comm.]. However, while approaches like these to the measurements of rates are important if predictions are to be made, extrapolation of rates from one situation to another must remain hazardous at least until the conditions can be defined and until their effect on rates of disappearance can be estimated.

3.1. *Potential Oxidative Capacity of Biotreaters*

Unfortunately it is not often easy to measure the rate of mineralization of one compound in a mixture. The rate of substrate stimulated respiration (oxygen uptake) can be converted to carbon removal and thus the potential oxidative capacity (P_{ox}) of a biotreater can be estimated given the volume and biomass concentration. Such measurements are simple to carry out and experiments with bench-scale systems have indicated that P_{ox} can be used in the analysis of operating problems.

The importance of viability measurements for biotreaters has been recognized for some time [Weddle and Jenkins, 1971; Blok, 1974] and attempts have been made to interpret such measurements in terms of the activity of the biocatalyst [Haas, 1979; Moira et al., 1979; Nelson and Lawrence, 1980]. The proposal is made here that the potential oxidative capacity (P_{ox}) should be used to measure this catalytic activity. The term is explained below and some illustrations are given where it has been applied to bench-scale biotreaters.

P_{ox} can be calculated as follows:

$P_{ox} = Q.A.V.X$, where Q = respiratory quotient [mg O_2 h^{-1} (g biomass)$^{-1}$] measured at maximum rate and corrected for endogenous respiration; A = a factor to convert oxygen uptake to removal of a carbon compound, conservatively taken to be about 0.4 for complete oxidation of most substrates; V = the volume (1) of the biotreater; and X = the biomass concentration (g dry wt l^{-1}).

This simple and approximate calculation has been used to predict the ability of bench-scale biotreaters to handle simulated spills. In each case where the P_{ox} was insufficient to deal with the new level of a spilled compound, problems in biotreater operation resulted.

As pointed out above, following a spill or the introduction of a new compound, some time is required for a mixed culture to respond and this response depends on the concentration of the compound in the reactor, which in itself would take some time to reach a new equilibrium for a fixed influent concentration in the absence of sufficient active biomass. Where degradation is possible the biomass responds by growth and/or increased enzyme synthesis. Thus P_{ox} provides a conservative estimate of the potential treatment rate.

The potential treatment rate may be widely divergent from the actual treatment rate and it is quite possible to arrive at erroneous conclusions about biotreater activity from measured treatment rate. For example, addition of nickel to a bench-scale biotreater led to a drop of 80% in the feed-stimulated rate without affecting performance of the unit. That is, the measured P_{ox} was still sufficient to account for the observed treatment rate.

Limitations to this approach include the partial oxidation of some chemicals to relatively more recalcitrant compounds. Here, P_{ox} could give a false measure of complete breakdown. However, experience in plant operation should allow this to be taken into account. Another limitation is the interpretation of the endogenous rate of respiration. This may represent the basic activity of the tricarboxylic acid cycle and associated respiratory activity in the biomass; however, it is unclear whether or not it should be included in estimation of P_{ox}. This can be allowed for in calculation of the factor A as can the assimilation of some carbon. Experience in measuring stoichiometry of oxygen uptake should allow a fairly accurate estimation of A.

Despite the limitations, the measurement of P_{ox} or some similar activity based on carbon removal, provides a useful approach to the estimation of catalytic activity. It is to be anticipated that many treatment systems operate below the maximum possible P_{ox} because of the operating conditions. In the calculation both Q, the respiration rate, and X, the biomass concentration, can be increased. It is important that the biomass is physiologically active and for a particular substrate (single compound or mixed) there is a limiting Q_{O_2} value such as would be obtained for a pure culture growing on a single compound at high dilution rate. For activated sludge systems operation at X values above 6 to 8 g solids l^{-1} is not generally considered feasible, although higher biomass concentrations may well be possible using different technologies such as fluidized beds. Assuming a maximum Q_{O_2} of about 200 mgO_2 h^{-1} (g biomass)$^{-1}$ and an X value of 8 g l^{-1} a maximum loading of 600 to 650 mg carbon l^{-1} h^{-1} is suggested.

4. Transients in Substrate Concentrations and Other Operating Parameters

Mixed cultures in biotreaters can be expected to respond to transients in the same way as pure cultures. Thus, the overall effect of temperature on activity of biomass from biotreaters treating domestic waste [Flegal and Schroeder, 1976] or chemical waste is essentially the same as that expected for most mesophilic microorganisms. As described above there should be little effect on the community within a CSTR-type biotreater as long as the activity is sufficient to maintain the concentration in the biotreater and effluent at a low level. The community can be expected to become unbalanced when excessive growth of primary utilizers results from increased carbon. If the concentration in the reactor increases significantly it is possible that different primary utilizers with faster growth rates will be selected [Veldkamp and Jannasch, 1972]. This provides a rational explanation for the appearance of bulking or dispersed bacteria following increases in carbon loading.

Suspended bacteria in the effluent from waste treatment systems are a major contributing factor to the BOD. This holds both for domestic wastes [Pipes, 1979] and chemical plants; in both cases the dispersed bacteria can give an erroneous impression of poor carbon removal.

Individual members of the communities may have responses that are sufficiently different to identify particular effects. For example, it is commonly observed that nitrification processes [Painter, 1970] are more sensitive than other conversions to changes in pH. It has not, however, been established whether these effects result from the potential capacity being closer to the actual treatment rate at the onset of the change.

It is possible that nitrite formed during incomplete nitrification could lead to other complications. Nitrite can react with secondary amines to produce N-nitrosamines some of which are suspected carcinogens. Some evidence has been produced to indicate that reaction between nitrite and secondary amines can be enzymatically catalysed [Kunisaki and Hayaishi, 1979]. Other transformations with similar potential consequences are possible. For example, Wyman et al. [1979] have shown that *Pseudomonas aeruginosa* converts 2,4,6-trinitrophenol (picric acid) to 2-amino-4,6-dinitrophenol (picramic acid) under conditions of oxygen depletion. Picramic acid but not picric acid was found to be directly mutagenic.

4.1. *Biotreatment of Phenol and Phenolic Wastes*

Carbonization effluents from coke-ovens represent one category of industrial effluent whose biotreatment has been studied in some detail. Such effluents are particularly interesting as they contain high concentrations of phenol and cyanide both of which are generally considered to be toxic to bacteria. Catchpole and Cooper [1972] found that degradation of components such as thiocyanate and cyanide was promoted by the addition of a number of organic compounds and they postulated that pyruvic acid played a key biochemical role in the improvement of the process. Jones and Carrington [1972] isolated eight strains from a bench-scale activated sludge plant which was treating carbonization effluent, two of which grew on phenol and one, probably *Thiobacillus thioparus*, that grew autotrophically on thiocyanate. Thiocyanate breakdown by the *Thiobacillus* species was unaffected by 1 mM phenol in pure culture but was completely inhibited in mixed culture with phenol and phenol-utilizers present. One of the phenol-utilizing strains showed Haldane-type inhibition by phenol with a K_i of about 1 mM.

The degradation of phenol by cultures in laboratory scale biotreaters has been investigated in further detail, including studies on anaerobic treatment [Healy and Young, 1978]. Yang and Humphrey [1975] also found a K_i of about 1 mM for inhibition of respiration by phenol but Beltrame et al. [1979] found, by measurements of the rate of phenol degradation *in situ*, that Monod kinetics were followed without inhibition.

Conflicting observations on inhibition by phenol may result from variation in the microbial population or from the different approaches to measurement of activity. Studies in the Shell laboratories have also indicated varying inhibition by phenol and have confirmed that phenol removal from a glycerol/phenol mixture is unlikely to be a problem under normal conditions. Biomass in a bench-scale biotreater adapted rapidly to a concentration of 2.8 mM phenol, both when the phenol concentration was raised from 0 to 2.8 mM and from 0.1 to 2.8 mM, with only a transient accumulation of phenol in the effluent. Respirometric measurement of P_{ox} allowed accurate prediction of the effluent phenol concentration following the spill and, with a sensitive radio-assay for phenol breakdown, indicated that, although phenol was inhibitory (50% inhibition at about 0.5 mM), some activity for phenol degradation was retained even after prolonged operation in the absence of phenol. The drop in activity following removal of phenol from the inflowing medium was, however, more rapid than could be accounted for by washout of active biomass. It is also interesting to note that, in this study, problems of dispersed growth and bulking were associated, respectively, with the enrichment in the clarifier supernatant of phenol- and glycerol-utilizing bacteria.

5. Separation of Biomass

In activated sludge systems, the separation of biomass can fail in two ways [Pipes, 1979]. If the settling velocity is less than the overflow rate, biomass will be removed from the upper part of the settling tank (clarifier). If the sludge compacts poorly, the return line does not recycle sludge and accumulated sludge eventually overflows into the effluent. These two extremes correspond to the phenomena of dispersed bacteria and bulking, respectively. It is difficult to generalize, as many variations exist in design of clarifying systems [Dick, 1976]. Interpretation of information on settling behaviour, thickening and separation of solids is complicated by the lack of reliable laboratory procedures for obtaining information which can be extrapolated to plant performance.

Why do microorganisms flocculate? Possibly the strongest selection for floc formation in waste treatment is that of survival in a recycle system. Those members of the community which do not settle are much more likely to be washed out. Flocculated organisms must retain activity during settling and recycling in order to retain a selection advantage. The environment within a floc may be variable and markedly different to that of microbes in free suspension. Unfortunately, little is known about the microenvironment within flocs. It can be speculated that affinity for substrates, including oxygen, is an important factor. At low substrate concentrations, it may be that the growth of filamentous forms is favoured by an ability to grow out along substrate gradients. The effect of process parameters on the development of filamentous forms has been studied recently by Houtmeyers *et al.* [1980] who concluded that several mechanisms probably contribute to floc formation, including some factors not directly related to the microorganisms [Busch and Stumm, 1968].

From the microbial aspect, the extent of filamentation and the amount and nature of extracellular polysaccharides are important. Bacterial strains selected for flocculent growth in pure culture possess fibrillar exocellular material [Friedman *et al.*, 1969]. As with the extent of filament formation too much polysaccharide may lead to poor settling [Harris and Mitchell, 1975]. Uronic acid residues in extracellular polymers have been implicated in the separation properties of activated sludge [Steiner *et al.*, 1976] and in the removal of metals [Brown and Lester, 1979]. Verachtert *et al.* [1980] have compared continuously and intermittently fed laboratory treatment systems using mixed media of glucose, starch and other carbohydrates. The continuously fed systems developed filamentous bacteria and bulking sludge [Houtmeyers *et al.*, 1980] in contrast to the intermittently fed system, where the sludge settled well. They explained the intermittent case by the floc-forming bacteria becoming dominant as a result of higher substrate uptake rates being able to survive starvation by utilizing intracellular polysaccharides. However, an alternative explanation in terms of concentration gradients within the flocs seems possible.

Forster [1968] concluded that metabolic processes can alter the surface of sludge particles and postulated that some physiological control of the settling properties of sludge might be possible. Unfortunately, his demonstration of a direct relationship between the sludge volume index and the electrophoretic mobility of sludge particles has not been further established. Conflicting results have been reported by Magara *et al.* [1976] who found that improved settling resulted from a reduction in both electrophoretic mobility and the amount of extracellular polysaccharide.

6. Possible Developments

A subjective view of the development of biotreatment is that it has suffered from its role as an unattractive requirement and from the philosophy that, where applied, waste treatment is an undesirable but necessary cost to society. Recent changes in public attitudes to the environment may result in a more positive approach to the economic aspects of waste treatment. This might not only lead to improved performance of waste treatment plants but also to improved performance of the industrial manufacturing plants, by pointing to the sources and quantities of particular wastes. Improved analysis of the streams involved is a first step to understanding performance;

in this connection the introduction of reliable instrumentation for determination of total organic carbon etc. will be of great use. For many industrial effluents more detailed analysis involving separation and identification of individual compounds will be required.

The most extensive application of waste treatment technology in recent years has undoubtedly taken place in the U.S.A. However, legislation has led to the installation of facilities on a time-scale which has largely precluded the development of new technology. A marvellous opportunity exists, as in many other areas of biotechnology, for the multidisciplinary development of solutions to some complex problems. For the microbiologist, one objective should be to understand the nature of the catalysts and other processes involved in waste degradation. Evidence has been discussed above that dead-end products and transient accumulation of intermediates can occur. For example, DiGeronimo et al. [1979] have shown that m-chlorobenzoate is converted to 4-chlorocatechol in sewage sludge and to 5-chlorosalicylic acid which is further metabolized. It seems probable that continuous exposure would result in selection of strains that are capable of more complete degradation of a greater range of seemingly recalcitrant compounds than is concluded from short-term experiments with unadapted biomass. Some of the mechanisms by which this could occur have been discussed by Slater and Godwin [1980] and the potential role of plasmids in evolution of new degradative capacity has been elegantly demonstrated by Knackmus and colleagues [Reineke and Knackmus, 1979; Hartmann et al., 1979]. One strain of Pseudomonas sp. was unable to utilize 2-chloro-, 4-chloro- or 3,5-dichlorobenzoic acids owing to the specificity of a dioxygenase which allowed growth on 3-chlorobenzoate. Another Pseudomonas sp. carried a plasmic coding for a benzoate dioxygenase which allowed conversion of 4-chloro- and 3,5-dichlorobenzoate. This latter strain could not grow on these compounds through an inability to degrade chlorocatechols. By a series of experimental challenges in the chemostat strains were selected which grew on the hitherto recalcitrant chlorobenzoates. Subsequently it was shown that genetic transfer had occurred.

Fall et al. [1979] tried to develop the capacity for degradation of 2,6-dimethyl-2-octene (DMO) in Pseudomonas citronellis. Although the attempted introduction of the OCT plasmid for octane degradation failed to generate strains which would grow on DMO, direct isolation of decane-utilizing strains growing on C_6 to C_{16} alkenes allowed growth on DMO which was degraded via citronellol. The wild type P. citronellis grew only on C_{12} to C_{16} alkanes.

These experiments and others which are discussed elsewhere [Slater and Somerville, 1979] illustrate that the microbial ecologist can already develop strains which will degrade seemingly recalcitrant compounds. For many industrial wastes some, at least, of the recalcitrant compounds may be relatively remote to existing metabolic pathways thus reducing the chances of natural selection of strains or communities with the requisite properties and also presenting a greater challenge to the experimental skills of microbiologists.

7. Acknowledgement

The author is grateful to his colleagues at Shell Development Co., Houston, for many useful discussions.

References

Banks, C.J., Davies, M., Walker, I. and Ward, R.D. (1976). Biological and physical characterization of activated sludge: a comparative experimental study at ten treatment plants. Water Pollution Control (1976), 492-508.

Banks, C.J. and Walker, I. (1977). Sonication of activated sludge flocs and the recovery of their bacteria on solids media. *Journal of General Microbiology* **98**, 363-368.

Beltrame, P., Beltrame, P.L., Carniti, P. and Pitea, D. (1979). Kinetics of phenol degradation by activated sludge: value of measurements in a batch reactor. *Water Research* **13**, 1305-1309.

Blok, J. (1974). Respriometric measurements on activated sludge. *Water Research* **8**, 11-18.

Boethling, R.S. and Alexander, M. (1979). Effect of concentration of organic chemicals on their biodegradation by natural microbial communities. *Applied and Environmental Microbiology* **37**, 1211-1216.

Brown, M.J. and Lester, J.N. (1979). Metal removal in activated sludge: the role of bacterial extracellular polymers. *Water Research* **13**, 817-837.

Bungay, H.R. and Bungay, M.L. (1968). Microbial interactions in continuous culture. *Advances in Applied Microbiology* **10**, 269-282.

Burns, R.G. (1979). Interaction of microorganisms, their substrates and their products with soil surfaces. In "Adhesion of Microorganisms to Surfaces" (eds. D.C. Ellwood, J. Melling and P. Rutter), pp.109-138. London: Academic Press.

Busch, P.L. and Sturnim, W. (1968). Chemical interactions in the aggregation of bacteria bioflocculation in waste treatment. *Environmental Science and Technology* **2**, 49-53.

Catchpole, J.R. and Cooper, R.L. (1972). The biological treatment of carbonization effluents III. New advances in the biochemical oxidation of liquid wastes. *Water Research* **6**, 1459-1474.

Clark, R.R., Chian, E.S.K. and Griffin, R.A. (1979). Degradation of polychlorinated biphenyls by mixed microbial cultures. *Applied and Environmental Microbiology* **37**, 680-685.

Dick, R.I. (1976). Folklore in the design of final settling tanks. *Journal of the Water Pollution Control Federation* **48**, 633.

DiGeronimo, M.J., Niklaido, M. and Alexander, M. (1979). Utilization of chlorobenzoates by microbial populations in sewage. *Applied and Environmental Microbiology* **37**, 619-625.

Downing, A.L., Tomlinson, T.G. and Truesdale, G.A. (1964). Effect of inhibitors on nitrification in the activated sludge process. *Journal Process Institute Sewage Purification*, 537-550.

Eikelboom, D.H. (1975). Filamentous organisms observed in activated sludge. *Water Research* **9**, 365-388.

Eikelboom, D.H. (1977). Identification of filamentous organisms in bulking activated sludge. *Progress in Water Technology* **8**, 153-161.

Fall, R.R., Brown, J.L. and Schaeffer, T.L. (1979). Enzyme recruitment allows the biodegradation of recalcitrant branched hydrocarbons by *Pseudomonas citronellis*. *Applied and Environmental Microbiology* **38**, 715-722.

Flegal, T.M. and Schroeder, E.D. (1976). Temperature effects on BOD stoichiometry and oxygen uptake. *Journal of the Water Pollution Control Federation* **48**, 2700-2706.

Forster, C.F. (1968). The surface of activated sludge particles in relation to their settling characteristics. *Water Research* **2**, 767-776.

Fredrickson, (1977). Behaviour of mixed cultures of microorganisms. *Annual Review of Microbiology* **31**, 63-87.

Friedman, B.A., Dugan, P.R., Pfister, R.M. and Remsen, C.C. (1969). Structure of extracellular polymers and their relationship to bacterial flocculation. *Journal of Bacteriology* **98**, 1328-1334.

Haas, C.N. (1979). Oxygen uptake rate as an activated sludge control parameter. *Journal of the Water Pollution Control Federation* **51**, 938-943.

Hacking, A.J., Aguilar, J. and Lin, E.C.C. (1978). Evolution of propanediol utilization in *Escherichia coli* mutant with improved substrate scavenging power. *Journal of Bacteriology* **136**, 522-530.

Harder, W., Kuenen, J.G. and Matin, A. (1977). A review: microbial selection in continuous culture. *Journal of Applied Bacteriology* **43**, 1-24.

Harris, R.H. and Mitchell, R. (1975). Inhibition of the flocculation of bacteria by biopolymers. *Water Research* **9**, 993-999.

Hartmann, J., Reineke, W. and Knackmus, H.J. (1979). Metabolism of 3-chloro-, 4-chloro- and 3,5-dichlorobenzoate by a pseudomonad. *Applied and Environmental Microbiology* **37**, 421-418.

Healy, J.B. and Young, L.Y. (1978). Catechol and phenol degradation by a methanogenic population of bacteria. *Applied and Environmental Microbiology* **35**, 216-218.

Houtmeyers, J., van den Eynde, E., Poffé, R. and Verachtert, H. (1980). Relations between substrate feeding pattern and development of filamentous bacteria in activated sludge processes. Part I. Influence of process parameters. *European Journal of Applied Microbiology and Biotechnology* **9**, 63-77.

Hughes, D.E. and Stafford, D.A. (1976). The microbiology of the activated sludge process. *Critical Reviews in Environmental Control*. September 1976, 233-257.

Jannasch, H.W. (1967). Growth of marine bacteria at limiting concentrations of organic carbon in seawater. *Limnology and Oceanography* **12**, 263-271.

Jones, J.G. and Carrington, E.G. (1972). Growth of pure and mixed cultures of microorganisms concerned in the treatment of carbonization waste liquors. *Journal of Applied Bacteriology* **35**, 395-404.

Kaspar, H.F. and Wuhrmann, K. (1978). Kinetic parameters and relative turnovers of some important catabolic reactions in digesting sludge. *Applied and Environmental Microbiology* **36**, 1-7.

Konings, W.N. and Veldkamp, H. (1980). Phenotypic responses to environmental change. In "Contemporary Microbial Ecology" (eds. D.C. Ellwood, J.N. Hedges, M.J. Latham, J.M. Lynch and J.H. Slater), pp.161-192. London: Academic Press.

Kunisaki, N. and Hayashi, M. (1979). Formation of N-nitrosamines from secondary amines and nitrite by resting cells of *Escherichia coli* B. *Applied and Environmental Microbiology* **37**, 279-282.

Magara, Y., Nambu, S. and Utosawa, K. (1976). Biochemical and physical properties of an activated sludge on settling characteristics. *Water Research* **10**, 71-77.

Mason, T.G. and Slater, J.H. (1979). Competition between an *Escherichia coli* tyrosine auxotroph and a prototrophic revertant in glucose- and tyrosine-limited chemostats. *Antonie van Leewenhoek* **45**, 253-263.

Moira, R., Dunn, I.J. and Bourne, J.R. (1979). Activated-sludge process dynamics with continuous total organic carbon and oxygen uptake measurements. *Biotechnology and Bioengineering* **21**, 1561-1577.

Nelson, P.D. and Lawrence, A.W. (1980). Microbial viability measurements and activated sludge kinetics. *Water Research* **14**, 217-225.

Oron, G., Shelef, G., Levi, A., Meydan, A. and Azov, Y. (1979). Algae/bacteria ratio in high-rate ponds. *Applied and Environmental Microbiology* **38**, 570-576.

Painter, H.A. (1970). A review of literature on inorganic nitrogen metabolism in microorganisms. *Water Research* **4**, 393-450.

Pipes, W.O. (1979). Bulking, deflocculation and pinpoint floc. *Journal of the Water Pollution Control Federation* **51**, 62-70.

Plimmer, J. (1978). In "Microbial Degradation of Pollutants in Marine Environments" (eds. E.D. Pritchard and A.W. Bourquin), Washington: United States Environmental Protection Agency.

Postgate, J.R. (1973). The viability of very slow-growing populations: a model for the natural ecosystem. *Bulletin of the Ecological Research Commission (Stockholm)* **17**, 237-242.

Reineke, W. and Knackmus, H.J. (1979). Construction of haloaromatic utilizing bacteria. *Nature* **277**, 385-386.

Sezgin, M., Jenkins, D. and Parker, D.S. (1978). A unified theory of filamentous activated sludge bulking. *Journal of the Water Pollution Control Federation*. 362-381.

Slater, J.H. (1978). The role of microbial communities in the natural environment. In "The Oil Industry and Microbial Ecosystems" (eds. K.W.A. Chater and H.J. Somerville), pp.137-154. London: Heyden and Sons.

Slater, J.H. and Godwin, D. (1980). Microbial adaptation and selection. In "Contemporary Microbial Ecology" (eds. D.C. Ellwood, J.N. Hedges, M.J. Latham, J.M. Lynch and J.H. Slater), pp.137-160. London: Academic Press.

Slater, J.H. and Somerville, H.J. (1979). Microbial aspects of wastes treatment with particular attention to the degradation of organic compounds. *Society for General Microbiology Symposium* **29**, 221-261.

Steiner, A.E., MacLaren, D.A. and Forster, C.F. (1976). The nature of activated sludge flocs. *Water Research* **10**, 25-30.

Stiff, M.J. (1978). Biodegradation of Surfactants. In "The Oil Industry and Microbial Ecosystems" (eds. K.W.A. Chater and H.J. Somerville), pp.118-128. London: Heyden and Sons.

Taber, W.A. (1976). Wastewater microbiology. *Annual Review of Microbiology* **30**, 263-277.

Veldkamp, H. and Jannasch, H.W. (1972). Mixed culture studies with the chemostat. *Journal of Applied Chemistry and Biotechnology (1972).* 105-123.

Verachtert, E., van den Eynden, E., Poffé, R. and Houtmeyers, J. (1980). Relations between substrate feeding pattern and development of filamentous bacteria in activated sludge processes. Part II. Influence of substrates present in the influent. *European Journal of Applied Microbiology and Biotechnology* **9**, 137-149.

Weddle, C.L. and Jenkins, D. (1971). The viability and activity of activated sludge. *Water Research* **5**, 621-640.

Wittauer, D.P. (1980). Biocoenosis and degradation in model wastewater treatment plants. *European Journal of Applied Microbiology and Biotechnology* **9**, 151-163.

Wyman, J.F., Guard, H.E., Wou, W.D. and Quay, J.H. (1979). Conversion of 2,4,6-trinitrophenol to a mutagen by *Pseudomonas aeruginosa*. *Applied and Environmental Microbiology* **37**, 222-226.

Yang, R.D. and Humphrey, A.E. (1975). Dynamic and steady state studies of phenol biodegradation in pure and mixed cultures. *Biotechnology and Bioengineering* **17**, 1211-1235.

Zamenhof, S. and Eichhorn, H.H. (1967). Study of microbial evolution through loss of biosynthetic functions: establishment of defective mutants. *Nature* **216**, 456-458.

Chapter 5

PROTOCOOPERATION OF YOGURT BACTERIA
IN CONTINUOUS CULTURES

F.M. DRIESSEN

Netherlands Institute for Dairy Research, Ede, The Netherlands

1. Introduction

In nature, mixed microbial cultures are common and the various species influence each other. Many natural phenomena may be only partially explained by studying pure cultures.
   The importance of mixed cultures in the dairy industry is well established. Cheese and butter manufacture, for example, involves inoculation with lactic streptococci and leuconostocs. To discuss all of this in one paper would be confusing, and it would in any case only be possible to enumerate some aspects of these particular mixed cultures without the appropriate background information. This paper is therefore restricted to one dairy product, namely yogurt. Our recent advances, especially in the field of the continuous manufacture of yogurt on a practical scale, will be discussed.
   Yogurt is made by the fermentation of milk by two types of bacteria, namely *Streptococcus thermophilus* and *Lactobacillus bulgaricus*. These two types of bacteria have in milk an interaction which is favourable to both, but not obligatory [Pette and Lolkema, 1950a]. Such a relationship is called a protocooperation [Odum, 1971].

2. Batch Cultures

   In a yogurt culture it is easy to follow the development of the two types of bacteria, because of the large differences in shape. They can be distinguished by direct microscopic count or by plate count techniques [Driessen *et al.*, 1977a; Galesloot *et al.*, 1961]. The yogurt system, therefore, lends itself to a fundamental study of the interaction between the two bacteria.
   In the mixed culture this interaction is seen very clearly in the acid production rate. The acid production rate of the mixed culture is much larger than the sum of

those of the corresponding pure cultures. This is shown
experimentally and the results are given in Table 1
[Pette and Lolkema, 1950a]. It appeared that the number

TABLE 1

*Acid Production and Number of Cells after an Incubation
for 3 h at 45°C of the Single and Mixed Cultures of*
Streptococcus Thermophilus *Strain Sts and* Lactobacillus
Bulgaricus *Strain Ib. [Pette and Lolkema, 1950*a].

Culture	Acid production (mmol ml^{-1})	Number of cells (x 10^6 organisms ml^{-1})	Kind of cells
4% S. thermophilus Sts	24	500	diplococcus
1% L. bulgaricus Ib	20	180	rod
4% S. thermophilus Sts		880	diplococcus
+	74		
1% L. bulgaricus Ib		170	rod

of diplococci in the mixture was larger than in the pure
culture, whilst the number of rods was almost the same in
both cases. For reasons given later it is emphasized
that strongly heated milk was probably used in this
investigation (see Section 2.2). It is assumed that the
higher acid production rate is a result of the better
growth of the cocci. The acid production per cell in a
batch culture of *Streptococcus thermophilus* as a pure
culture and mixed with *Lactobacillus bulgaricus* depends
on the growth rate of this bacterium [Pette and Lolkema,
1950a].

2.1. *Stimulation of the Cocci*

The growth curve of a pure culture of *Streptococcus
thermophilus* declines at an early stage. *Lactobacillus
bulgaricus* was capable of extending the exponential
growing phase of the cocci. Therefore it was concluded
that milk contained insufficient compounds necessary for
good growth of the cocci. The production of these compounds in milk by the lactobacilli was first investigated
by Pette and Lolkema [1950a,b].

By using a bacteria-free filtrate of *L. bulgaricus*
culture in milk it was proved that the stimulation of
S. thermophilus was due to the presence of some water-soluble nutrients released by the *L. bulgaricus* strains
[Pette and Lolkema, 1950b; Moon and Reinbold, 1976].
Further it appeared that yeast autolysate had a similar
effect on *S. thermophilus*. While amino acids are

available in yeast autolysate and it had been known that
L. *bulgaricus* could liberate amino acids from casein
[Orla-Jensen, 1919], it was thought that the stimulus
consisted of those amino acids which are usually scarce
in normal milk. Milk itself contains very few free amino
acids. The total amount of small peptides and amino
acids in milk is about 50 mg kg^{-1} [Walstra and Ven der
Haven, 1979], and Pette and Lolkema [1950b] could not
detect free amino acids in sterilized milk. This assumption was investigated by various research workers, and
their results were not identical (Table 2). Possible

TABLE 2

Survey of the Effect of Amino Acids on the Acid Production of Strains of Streptococcus Thermophilus

Authors	Medium	Mixture of 18 amino acids	
		Reduced growth when omitted	Necessary for optimal activation
Pette and Lolkema [1950b]	milk	Asp, Cys, Val, His, Lys, Leu (in Autumn: Gly, Ile, Tyr, Glu, Met)	Val
Bautista et al. [1966]	milk	Gly, His	Gly, His
Accolas et al. [1971]	milk	Val, His, Leu	Val, His, Leu
Desmazeaud [1974]	milk	Val, His, Met	Val, His, Met
Higashio et al. [1977a]	milk	Val, His, Met, Glu, Leu	Val
Shankar and Davies [1977]	milk	Val, His, Met, Glu, Leu, Try	Val, His, Met, Glu, Leu, Try
Bracquart et al. [1978]	milk	Val, His, Met, Glu, Leu, Try, Cys, Ile, Tyr Arg	His, Met, Glu
Nurmikko and Kärhä [1963]	synthetic	Val, His, Met Glu, Leu, Try, Cys, Thr, Gly	Glu, Cys, Try

explanations for the variations in the demand of amino
acids found by the different investigators could be

seasonal effects or a heavy bacterial growth in milk before the heat treatment, by which the casein could be partly hydrolysed. After inoculation of the milk with a yogurt culture L. bulgaricus stimulates the growth of S. thermophilus by liberating the essential amino acids and peptides from the casein [Miller and Kandler, 1964; Shankar and Davies, 1978].

2.2. Stimulation of the Lactobacilli

The acid production of Lactobacillus bulgaricus is stimulated by compounds produced by Streptococcus thermophilus in the absence of oxygen or at low oxygen tension. Accelerated acid production is not observed in cultures with high oxygen tensions [Galesloot et al., 1968]. This effect is shown in Table 3. S. thermophilus cultures at different stages of early growth were cultivated under an atmosphere of nitrogen or air, then neutralized to pH 6.6 and pasteurized at 80°C for 10 min. They were then inoculated with L. bulgaricus and incubated.

TABLE 3

Souring of Lactobacillus Bulgaricus Strain Ib in Neutralized and Pasteurized Skim Milk Cultures of Streptococcus Thermophilus Strain Sts Grown Under Air or Under Nitrogen in Flasks Filled Halfway with Skim Milk. Skim Milk Heated at 90°C for 10 min. [Galesloot et al., 1968].

S. thermophilus culture		Souring of L. bulgaricus (mmol ml^{-1})	
pH before treatment	Atmosphere	Treated S. thermophilus culture	Control
6.1	air	31	31
6.1	nitrogen	56	
5.9	air	32	30
5.9	nitrogen	62	
5.7	air	33	31
5.7	nitrogen	67	

Orla-Jensen and Jacobsen [1930] showed that the acid production of L. bulgaricus is accelerated in strongly heated milk. From later investigations it is known that sodium formate stimulates the growth of thermophilic lactobacilli in mildly heated milk, and that by autoclaving or steaming the milk, formate is liberated from the lactose [Auclair and Portman, 1959; Galesloot et al., 1968]. For this latter reason it was a long time before it was shown that the lactobacilli are stimulated by the streptococci. It also explains why Pette and Lolkema did not find stimulation of lactobacilli in the mixed culture using strongly heated milk (Table 1).

Veringa et al. [1968] proved that the stimulating compound for the lactobacilli, produced by the cocci, was formic acid. The volatile fatty acids from a culture of S. thermophilus strain Sts were determined by column chromatography. The results are given in Fig. 1. The

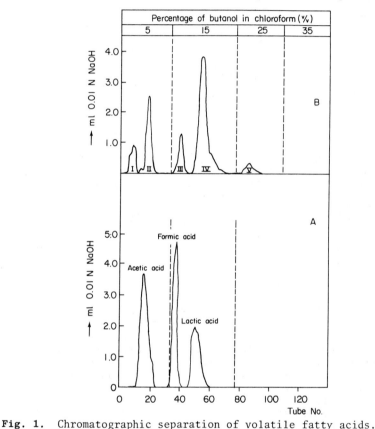

Fig. 1. Chromatographic separation of volatile fatty acids.
A. Mixture of acetic, formic and lactic acid.
B. Acids isolated from a skim milk culture of Streptococcus thermophilus strain Sts.
[Veringa et al., 1968].

separated peaks I to V were tested for stimulating activity towards the growth of L. bulgaricus strain Ib in mildly heated skim milk. The test for stimulation was carried out with the solution obtained by dissolving each fraction. The results are given in Table 4 [Veringa et al., 1968]. Only the fraction containing formic acid was found to have a stimulating activity on the L. bulgaricus

TABLE 4

*Stimulating Activity of the Individual Fatty Acids,
Isolated from a Skim Milk Culture of* Streptococcus Thermophilus *Sts,
on the Souring of* Lactobacillus Bulgaricus *Ib.
Skim Milk Heated at 90°C for 10 min. [Veringa et al., 1968].*

Peak number	Identified as	Acidity (°N)
I	higher fatty acids	28
II	acetic acid	28
III	formic acid	66
IV	lactic acid	36
V	not identified	31

in mildly heated milk. The small activity of peak IV was attributed to some formic acid caused by incomplete separation. That formate is an important stimulant was confirmed and quantified by other investigators [Accolas et al., 1971; Higashio et al., 1977b; Shankar and Davies, 1978]. The heat treatment used in practice for preparing milk for yogurt making, for instance 10 min. at 85°C, is such that *L. bulgaricus* strongly needs the formate produced by *S. thermophilus*.

3. Continuous Cultures

In a continuous culture of bacteria the dilution rate (D) of the bacteria equals the specific growth rate (μ) at steady state conditions. In a continuous culture of yogurt bacteria the growth rate of *Streptococcus thermophilus* and of *Lactobacillus bulgaricus* must be the same and equal to the dilution rate ($\mu_{Sc} = \mu_{Lb} = D$). This would be a very exceptional situation, but these cultures appeared to be very stable. According to the theory of the chemostat culture one of the organisms must either be washed from the growth vessel or completely replace the other species in the chemostat culture [Tempest, 1970]. No changes in the bacterial balance could be observed at each of the pH values tested [Girginov, 1965; Driessen et al., 1977a; MacBean et al., 1979].

3.1. *Models for Protocooperation*

In yogurt we are dealing with a system of two microbial populations having a complementary metabolism. This means that each population produces a substance — not present in the initial or feed medium — capable of stimulating the growth of the other organism. This is rather a complicated system. A model for such a protocooperation was developed by Wilkinson et al. [1974] in which a substrate-limited and product-inhibited situation is described. The dynamics of a complementary system such as is present in yogurt have been further worked out by means of mathematical models of growth [Meyer, 1975; Meyer et al., 1975]. The interaction is outlined schematically in Fig. 2. The mass balances of the bacteria in

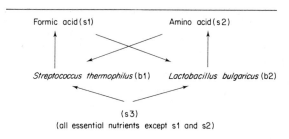

Fig. 2. Scheme of complementary metabolism of two microbial populations [Meyer et al., 1975].

this scheme are as follows:

increase = -dilution + growth

or $db_1/dt = -D\ b_1 + \mu\ b_1$
and $db_2/dt = -D\ b_2 + \mu\ b_2$

and the substrate balances:

increase = -dilution - consumption + production

or $ds_1/dt = -D\ s_1 - \alpha\ \mu\ b_1 + \gamma\ \mu\ b_2$
and $ds_2/dt = -D\ s_2 - \delta\ \mu\ b_2 + \beta\ \mu\ b_1$

(α, β, γ and δ are stoichiometric constants)

This system can only exist if:

production > consumption

or $\beta\gamma > \alpha\delta$

and $D < \mu_{max\ 1,2}$, of each of the organisms in the vessel.

This model is developed further for competitive and inhibitive substrates, for the system of Meyer *et al.* [1975] as such is not self-correcting. For any small disturbance of the original condition the inevitable consequence is a still greater change in the same direction, which makes it unstable. For instance, a decrease of the amino acid concentration is followed by a smaller growth rate of *Streptococcus thermophilus* and consequently a decreased production of formic acid and the culture is washed out [Meyer, 1975; Meyer *et al.*, 1975].

What is needed in the given scheme is an extra inhibiting factor to prevent an uncontrolled situation. This could possibly be, in yogurt, an inhibiting product, such as lactic acid. For the purposes of argument it is not of importance whether the *Lactobacillus* or the *Streptococcus* is inhibited. The new scheme is presented in Fig. 3 [Ubbels, J. pers. comm.].

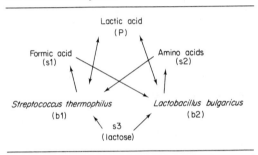

Fig. 3. Scheme of complementary metabolism of two microbial populations with product inhibition.

How does this product inhibition function in this system? The best way to demonstrate this is in a three-dimensional figure (see Fig. 4). The growth rates of both bacteria in relation to the substrate concentrations are given in planes. The influence of the product inhibition is visualized by a deviation of the ideal Monod curve. At higher lactic acid concentrations the growth rate will diminish. The planes intersect each other. The intersecting line symbolizes the points at which the steady states are reached. This intersecting line has an optimum, caused by the product inhibition.

At the intersecting line of area II, it is seen that the growth rate decreases with increasing substrate concentration due to the increased product formation. At the lower growth rate less product is formed, and growth rate is, therefore, stimulated again. This means for area II that when the culture is maintained at a particular pH, or a particular lactic acid concentration, there will be a steady state.

When we look at the intersecting line area I, we see

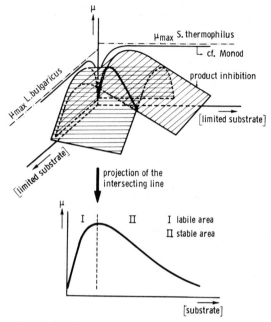

Fig. 4. Three-dimensional graph of the complementary metabolism of yogurt bacteria with product inhibition. Steady states are only possible at the intersecting line.

that an increased substrate concentration is followed by an increasing growth rate, and a decreased substrate concentration by a decreasing growth rate. In other words: area I is not stable.

3.2. *Behaviour of the Mixed Cultures*

The growth rate of the yogurt bacteria was investigated, together and separately, in a pH controlled fermentor. The rate of the feed was controlled to maintain a constant pH, that is a constant lactic acid concentration. Assuming a linear relationship between acid produced and cell mass, the pH-stat was considered to operate in a similar manner to a turbistat with the organism growing substrate unlimited, although end product inhibited, and thus at the maximum specific growth rate for the set-point pH and product concentration. Dealing with milk as the feed, the choice of growth rate limiting parameters is restricted.

It was found experimentally that the dilution rate decreases at decreasing pH values [Driessen et al., 1977a]. These results are given in Fig. 5. From this plot it can be seen how the dilution rate increases at increasing pH values. It appears from Fig. 5 that at a pH value as

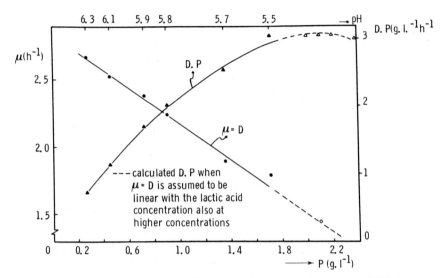

Fig. 5. Relationship between the specific growth rate μ (which equals the dilution rate D), lactic acid production rate (DP) of yogurt bacteria, and the lactic acid concentration (P) in the medium [Driessen et al., 1977a].

high as 6.3 the culture is in area II (compare with Fig. 4). This is in accordance with earlier observations with lactic acid bacteria [Rogers and Whittier, 1928; Leudeking and Piret, 1959; Linklater and Griffin, 1971; Petterson, 1975].

There is a linear relationship between the concentration of lactic acid (P) and the specific growth rate (μ) in the limited range of experimental conditions [Keller and Gerhardt, 1975; Driessen et al., 1977a]. From these results it appeared also that area I of the intersecting line is situated at low substrate concentrations or close to the pH of milk.

It is important to realize that after inoculation with the bacteria the culture is growing batchwise through the unstable area I. After that a steady state can be reached.

3.3. *Stimulation of* Streptococcus Thermophilus

In Fig. 5 the relationship is shown between the dilution rate of the yogurt culture and the lactic acid concentration. The question was investigated as to whether the relationship was dependent on a limiting substrate. Therefore the yogurt culture was grown in steady state conditions during 16 h in the pH range of 5.5 to 6.3 and then either enzymatically hydrolysed casein (2.0 g l^{-1}) or lactose (20 g l^{-1}) were added to the feed. The addition of lactose had no effect on the dilution rate, the numbers of diplococci and rods or their ratio. These characteristics remained constant. Therefore it was concluded that the continuous culture of this particular yogurt culture ISt was not limited by this substrate.

On the other hand the dilution rate was enhanced by the addition of the hydrolysed casein. Results of an experiment at pH 5.9 are given in Fig. 6. The addition of the hydrolysed casein probably affected the pH of the culture. By this interference the dilution rate was disturbed temporarily. In this mixed culture the growth of

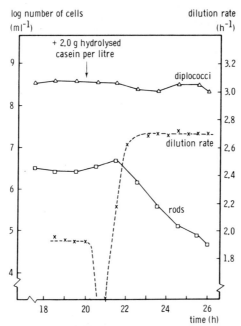

Fig. 6. Effect of enzymatically hydrolysed casein (2.0 g l^{-1}) on the yogurt bacteria ISt in continuous culture in skim milk at pH 5.9.

the *S. thermophilus* was substrate limited. When the hydrolysed casein was added to the culture, the diplococci were no longer inhibited by the lack of substrate. The dilution rate increased considerably. Attention had to be paid to the very short generation time of the streptococci; in this situation the generation time decreased from 21.6 to 15.4 minutes. The generation time of *Lactobacillus bulgaricus* remained the same, and this organism was washed out.

The results are not contrary to those of MacBean *et al.* [1979], who predicted and confirmed by experiment that the mixed culture of *S. thermophilus* TS2 and *L. bulgaricus* LB1 in a pH-stat (5.5) continuous yogurt fermentor was stable. The growth of *L. bulgaricus* LB1 was not substrate limited (D = μ_{max} for *L. bulgaricus* LB1). The growth of *S. thermophilus* TS2, however, became substrate limited so that its growth rate equalled the existing dilution rate.

A continuous culture of pure *S. thermophilus* strain Sts in milk showed a dilution rate of only 0.6 h^{-1} at pH 5.8 (Fig. 7). After the addition of the hydrolysed casein the dilution rate increased to 2.2 h^{-1}, whilst the number of diplococci remained constant and at the same level as in the mixed culture. This is shown in Fig. 7.

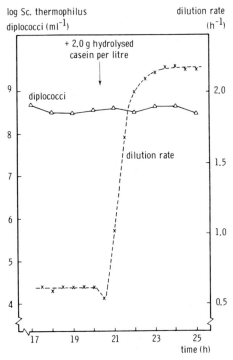

Fig. 7. Effect of enzymatically hydrolysed casein (2.0 g l^{-1}) on the growth and acid production of *Streptococcus thermophilus* Sts in continuous culture in skim milk at pH 5.8.

The production rate of the organisms and the lactic acid production rate are enhanced by the addition of hydrolysed casein to the feed. A difference exists between the dilution rate of the streptococci in the mixed culture and in the pure culture (2.7 h^{-1} and 2.2 h^{-1} respectively). The pH of the mixed culture was higher, which caused a higher dilution rate. Besides that the hydrolysed casein probably differs from the amino acid and small peptides produced by *L. bulgaricus* in the mixed culture. There may also be some differences in the composition of the milk, used in the two experiments.

It is clear from the low dilution rate of *S. thermophilus* Sts in milk that in the mixed culture the hydrolysed casein is necessary for rapid growth. The proteolytic activity of the lactobacilli, therefore, stimulates the growth of *S. thermophilus* Sts.

3.4. *Stimulation of* Lactobacillus Bulgaricus

The addition of sodium formate (0.020 g l^{-1}) to a yogurt culture at steady state had no effect. The effect of formic acid on the growth of *L. bulgaricus* Ib in a continuous culture in milk was investigated at pH 6.0. The dilution rate of the culture was about 0.2 h^{-1}. After addition of sodium formate (0.02 g l^{-1}) a new steady state was reached at a higher level. This is shown in Fig. 8.
In contrast to the previous experiment with *Streptococcus thermophilus* the number of *L. bulgaricus* in a pure culture was about five times higher than in the mixed culture. After the addition of the sodium formate (0.020 or 0.040 g l^{-1}) this number increased seven fold. The dilution rate increased too, about six times. The increased dilution rate can be explained arithmetically from the growth of *L. bulgaricus*, whilst the addition of the substrate in the pure culture of *S. thermophilus* caused an additional increased lactic acid production rate.

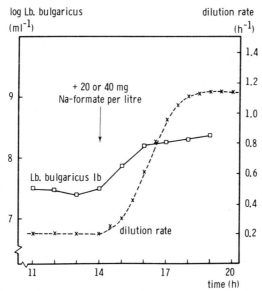

Fig. 8. Effect of sodium formate (0.02 and 0.04 g l^{-1}) on the growth and acid production of *Lactobacillus bulgaricus* Ib in continuous culture in skim milk at pH 6.0.

After the addition of the sodium formate, the maximum dilution rate of the lactobacilli alone was about half that of the mixed culture. The same results were obtained when the feed consisted of milk in which *S. thermophilus* had been grown to pH 6.0, successively neutralized and pasteurized. The formic acid content in this feed was estimated according to Lang and Lang [1972] after the milk was clarified [Carrez, 1909]. When the dilution rate of *S. thermophilus* Sts was controlled to pH 6.0 the content of sodium formate in the milk was increased to 0.032 g l^{-1}. This was in accordance with the amount found experimentally by Galesloot *et al.* [1968], but more than was necessary with an artificial increase, in which 0.02 g l^{-1} appeared to be sufficient. The production of sodium formate by *S. thermophilus* strain Sts was in excess of the need of *L. bulgaricus* strain Ib.

It might be possible that the higher number of lactobacilli in the pure culture than in the mixed culture was caused by the fact that in the mixed culture both bacteria co-operatively provide the required acid production. The number of *L. bulgaricus* Ib in the mixed culture was just high enough to provide the remaining acid required for the steady state of that culture. If this was true, there had to be a second steady state, because the dilution rate of the mixed culture could not be obtained in a continuous culture of pure *L. bulgaricus* Ib. The number of lactobacilli in the mixed culture was lower than in the pure culture. The possibility of another steady state for the pure culture was investigated, with milk to which sodium formate (0.020 g l^{-1}) was added. After reaching the steady state the dilution rate was increased to 90% of that of the mixed culture, and the pH was kept constant by addition of lactic acid. The results are given in Fig. 9. In this figure it is clearly shown that under these circumstances the *L. bulgaricus* Ib is washed out and there is no second steady state. Since the experimental conditions in this investigation were the same for all the continuous cultures, there is no support for the assumption that the number of lactobacilli in the mixed culture is determined by wall growth. Under wash-out conditions the number of lactobacilli decreases until far below the number in the steady state of the mixed culture. Besides, as already stated, the streptococci strongly need the amino acids and peptides produced by the lactobacilli. This is firm evidence that wall growth in the fermentor does not play a role.

It is concluded from these experiments that formic acid is not the only limiting substrate for *L. bulgaricus* in the continuous culture of the yogurt bacteria.

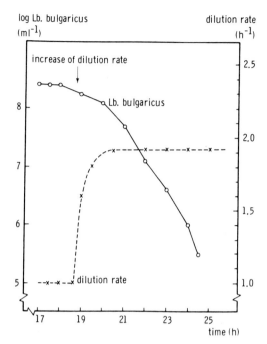

Fig. 9. Effect of the increase of the dilution rate of a continuous culture of *Lactobacillus bulgaricus* Ib in skim milk with sodium formate (0.02 g l^{-1}) at pH 5.9, to 90% of the dilution rate of the mixed yogurt culture. The pH is kept at a constant value by the addition of 1M lactic acid.

3.5. *Mixed Culture under Aerobic Conditions*

The previous assumption can be checked fairly easily. *Streptococcus thermophilus* does not produce formic acid when the culture is kept under aerobic conditions [Galesloot et al., 1968] (Table 3). This phenomenon was used during the further investigation. When the mixed culture was cultivated under aerobic conditions, there was a good chance that only one characteristic for optimal protocooperation would fail, namely the formic acid. The unknown substance would be present in the culture. The results of such an experiment are shown in Fig. 10. After the addition of the sodium formate the dilution rate increased very rapidly, and a new steady state was reached (D = 1.8 h^{-1}) at a value very close to that of the mixed culture under anaerobic conditions (D = 1.9 h^{-1}). There were two striking differences in the effect of the addition of the formate on *Lactobacillus bulgaricus* in a continuous pure culture and in a continuous mixed culture under anaerobic conditions. In the first situation there was only an increased growth rate, whilst in the latter situation there was an increased growth rate and an

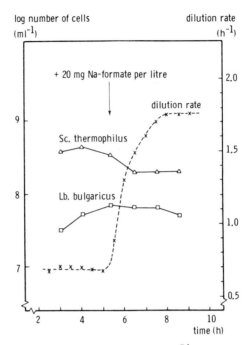

Fig. 10. Effect of sodium formate (0.02 g l^{-1}) on the dilution rate of a continuous yogurt culture RR in skim milk under aerobic conditions at pH 5.7.

increased production rate. The dilution rate in the latter situation was much higher. These facts support the assumption that formic acid is not the only limited substrate for *L. bulgaricus* produced by *S. thermophilus*.

3.6. *Another Stimulus for* Lactobacillus Bulgaricus

According to Japanese investigators pyruvic acid is also able to stimulate *L. bulgaricus*. The pyruvic acid could be replaced by other organic acids with the same effect, but formic acid is needed [Higashio *et al.*, 1977b]. We tried to confirm this observation but we did not succeed. The observation that the stimulating effect of pyruvic acid could be replaced by other organic acids may be connected with the fact that lactic acid bacteria are able to produce pyruvic acid from these compounds [Radler, 1975].

There are some indications that the other limiting substrate is possibly a volatile compound. It is known that yogurt bacteria can produce carbon dioxide [Abrahamsen, 1977; Flückinger and Walser, 1977], while *Streptococcus thermophilus* B produced carbon dioxide during growth in milk [Hup and Stadhouders, 1979]. In the meantime it was found that carbon dioxide is able to stimulate to a less extent the growth of L. *bulgaricus* in continuous culture.

The role of carbon dioxide in the metabolic pathways of bacteria is reviewed by Wood and Stjernholm [1962]. We have indications that the continuous culture of L. *bulgaricus* cultivated in milk with sodium formate and carbon dioxide can reach a dilution rate as high as that of the mixed culture under the same conditions.

3.7. Continuous Manufacture of Yogurt

Based on the stability of this mixed culture, several investigators have been trying to develop a procedure for the continuous manufacture of yogurt [Girginov, 1965; Lelieveld, 1976; Driessen et al., 1977a,b; MacBean et al., 1979]. The research was started because of the increased consumption of the stirred type of yogurt. This type of yogurt is cultivated in a tank, stirred, cooled and packaged. Dairy factories had to increase their production. By this scaling up, problems rose with regard to process control. It is often impossible to package and cool a large production in a suitable time, so that the yogurt becomes too sour and the consistency is spoiled by the rapid filling. A continuous process might offer a solution for these problems.

The continuous manufacture of yogurt is based upon a two-stage process. The first stage consists of a pre-fermentation of the milk. This stage is performed at 45°C, the optimum temperature for growth of yogurt bacteria and limited to a pH of 5.7 because of the occurrence of syneresis below that pH value [Girginov, 1965; Driessen et al., 1977a,b]. The production of bacteria takes place in the first stage. The pH is maintained at a constant level by neutralizing the lactic acid production with fresh yogurt milk. The dilution rate of the pre-fermentation tank is about 1.9 to 2.0 h^{-1}. The bacteriological aspects of this pre-fermentation have been discussed in the foregoing.

In the second stage the formation of the yogurt texture takes place, caused by the further acidification. At the end of this stage the yogurt is stirred. A schematic diagram of the continuous manufacture of yogurt is given in Fig. 11 [Driessen et al., 1977b].

The pre-fermented milk is transported to the coagulation tank. To avoid any unwanted mixing of the transferred milk, this milk is distributed smoothly on the surface of the acidifying milk in the coagulation tank. A solution of this problem may be obtained by the use of a centrifugal distributor [Dutch patent, 1975a].

At the early stages of the coagulation the milk must not be disturbed, otherwise the yogurt will start wheying off. This is only possible if the acidifying milk passes through the coagulation tank in a plug flow. A plug flow is characterized by a flat profile of streaming. In

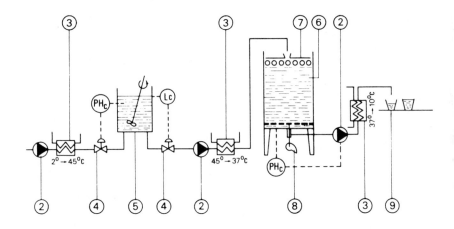

Fig. 11. Schematic diagram of the continuous manufacture of yogurt. 1. Bulk of yogurt milk; 2. pump; 3. heat exchanger; 4. automatically controlled valve; 5. prefermentation vessel; 6. coagulation tank; 7. centrifugal distributor; 8. stirring plate; 9. filling equipment [Driessen et al., 1977b].

flowing through the tank a gradient of acidity will arise, and the texture of the yogurt is built up in proportion to this gradient.

To start the process the coagulation tank is filled until a certain level is reached. This level is kept at a constant value. By the formation of lactic acid the pH will decrease, which causes the coagulation of the casein in the milk. This is the basis of the yogurt texture and this process is proportional to the residence time of the fermenting milk in the coagulation tank. In general a pH value of 4.3 is judged to be the best for yogurt. The yogurt is stirred with a specially designed plate, which allows a stirring treatment in the coagulation tank without any disturbance of the coagulating milk [Dutch patent, 1975b]. The total time from milk to yogurt is about 3 h. By this procedure the acidity and the viscosity of the final yogurt can be controlled between certain limits. The organoleptic characteristics of the continuous manufactured yogurt are good.

An impression of the equipment for the production of 4000 l h^{-1} is given in Fig. 12. By the realization of this new technology an enormous space saving is obtained. In comparison with the traditional manufacture of yogurt, only 20% of the tank volume is needed for the same production capacity.

Fig. 12. Installation for the continuous manufacture of yogurt with a capacity of 4000 l h^{-1} [Stork, Amsterdam]. This apparatus is assembled at the Netherlands Institute for Dairy Research.

Acknowledgements

The author wishes to thank Mr J. Stadhouders and Mr J. Ubbels for their kind comments and Mr F. Kingma for his technical assistance.

References

Abrahamsen, R.K. (1977). Changes in some bacteriological and biochemical activities in yogurt starters during seven hours incubation. *Mejeriposten* **66**, 603-611; 639-646.

Accolas, J.P., Veaux, M. and Auclair, J. (1971). Etude des interactions entre diverses bactéries lactiques, thermophiles et mésophiles, en relation avec la fabrication des fromages à pâte cuite. *Le Lait* **51**, 249-272.

Auclair, J.E. and Portman, A. (1959). Effet stimulant du lait autoclavé sur la croissance de *Lactobacillus lactis* role de l'acide formique. *Le Lait* **39**, 496-519.

Bautista, E.S., Dahiya, R.S. and Speck, M.L. (1966). Identification of compounds causing symbiotic growth of *Streptococcus thermophilus* and *Lactobacillus bulgaricus* in milk. *Journal of Dairy Research* **33**, 299-307.

Bracquart, P., Lorient, D. and Alais, C. (1978). Effet des acides amines sur la croissance de *Streptococcus thermophilus*. II- Etude sur cinq souches. *Milchwissenschaft* **33**, 341-344.

Carrez, C. (1909). "Handbuch der Lebensmittelchemie" pp.331-332. Berlin, Heidelberg, New York: Springer Verlag.

Desmazeaud, M. (1974). Etude des peptides stimulant la croissance de *Streptococcus thermophilus* sur le lait. Thesis, University of Caen, France.

Driessen, F.M., Ubbels, J. and Stadhouders, J. (1977a). Continuous manufacture of yogurt. I. Optimal conditions and kinetics of the prefermentation process. *Biotechnology and Bioengineering* **19**, 821-839.

Driessen, F.M., Ubbels, J. and Stadhouders, J. (1977b). Continuous manufacture of yogurt. II. Procedure and apparatus for continuous coagulation. *Biotechnology and Bioengineering* **19**, 841-851.

Dutch patent (1975a). Inrichting voor het vrijwel zonder vermengen toevoeren van een vloeistof aan een vloeistofoppervlak Nr. 75.09155.

Dutch patent (1975b). Inrichting voor het teweegbrengen van afschuifkrachten bij de behandeling van yoghurt en andere viskeuze vloeistoffen of gelen. Nr. 75.08982.

Flükiger, E. and Walser, F. (1977). Beitrag zur Kenntis der CO_2-Bildung von Rohmischkulturen. *Schweizerische Milchzeitung* **103**, 640.

Galesloot, T.E., Hassing, F. and Stadhouders, J. (1961). Agar media voor het isoleren en tellen van aromabacteriën in zuursels. *Netherlands Milk and Dairy Journal* **15**, 127-150.

Galesloot, T.E., Hassing, F. and Veringa, H.A. (1968). Symbiosis in yogurt (I). Stimulation of *Lactobacillus bulgaricus* by a factor produced by *Streptococcus thermophilus*. *Netherlands Milk and Dairy Journal* **22**, 50-63.

Girginov, T. (1965). Bulgarische Sauermilch (Joghurt) and die Methoden der Herstellung. *Lebensmittel Industrie* **12**, 263-266.

Higashio, K., Yoshioka, Y. and Kikuchi, T. (1977a). Isolation and identification of growth factor of *Streptococcus thermophilus* produced by *Lactobacillus bulgaricus*. *Journal of the Agricultural Chemical Society of Japan* **51**, 203-208.

Higashio, K., Yoshioka, Y. and Kikuchi, T. (1977b). Isolation and identification of growth factor of *Lactobacillus bulgaricus* produced by *Streptococcus thermophilus*. *Journal of the Agricultural Chemical Society of Japan* **51**, 209-215.

Hup, G. and Stadhouders, J. (1979). Growth of thermoresistant streptococci in cheese-milk pasteurizers. 3. Specific bacterial flora and its effect on the quality of cheese. *Zuivelzicht* **71**, 1141-1143.

Keller, A.K. and Gerhardt, P. (1975). Continuous lactic acid fermentation of whey to produce a ruminant feed supplement high in crude protein. *Biotechnology and Bioengineering* **17**, 997-1018.

Lang, E. and Lang, H. (1972). Spezifische Farbreaktion zum direkten Nachweis der Ameisensäure. *Zeitschrift fuer Analytische Chemie* **260**, 8-10.

Lelieveld, H.L.M. (1976). Continuous fermentation in yogurt

manufacture. *Process Biochemistry* **11**, 39-40.
Linklater, P.M. and Griffin, C.J. (1971). Growth of *Streptococcus lactis* in milk in a continuous fermenter. *Journal of Dairy Research* **38**, 127-144.
Leudeking, R. and Piret, E.L. (1959). A kinetic study of the lactic acid fermentation. Batch process at controlled pH. *Journal of Biochemical and Microbiological Technology and Engineering* **1**, 393-412.
MacBean, R.D., Hall, R.J. and Linklater, P.M. (1979). Analysis of pH-stat continuous cultivation and the stability of the mixed fermentation in continuous yogurt production. *Biotechnology and Bioengineering* **21**, 1517-1541.
Miller, I. and Kandler, O. (1964). Untersuchungen über den Eiweissabbau in Sauermilchen I. Mitteilung: Die Freien Aminosäuren in Joghurt, Bioghurt and Acidophilusmilch. *Medizin und Ernährung* **5**, 100-108.
Meyer, J.S. (1975). Dynamics of mixed populations having complementary metabolisms. Thesis University of Minnesota.
Meyer, J.S., Tsuchiya, H.M. and Fredrickson, A.G. (1975). Dynamics of mixed populations having complementary metabolism. *Biotechnology and Bioengineering* **17**, 1065-1081.
Moon, N.J. and Reinbold, G.W. (1976). Commensalism and competition in mixed cultures of *Lactobacillus bulgaricus* and *Streptococcus thermophilus*. *Journal of Milk and Food Technology* **39**, 337-341.
Nurmikko, V. and Kärhä, E. (1963). Nutritional requirements of lactic acid bacteria. II. Vitamin and amino acid requirements of *Streptococcus thermophilus* strains. *Dairy Science Abstracts* **25**, 2034.
Odum, E.P. (1971). "Fundamentals of Ecology", pp.211-212. Philadelphia, London, Toronto: W.B. Saunders.
Orla-Jensen, S. (1919). The lactic acid bacteria. Copenhagen, 78.
Orla-Jensen, S. and Jacobsen, J. (1930). Neue Untersuchungen über die bakteriziden Eigenschaften der Milch. *Zentralblatt fuer Bakteriologie*. II. **80**, 321-341.
Pette, J.W. and Lolkema, H. (1950a). Yoghurt I. Symbiose en antibiose in mengcultures van *Lb. bulgaricus* en *Sc. thermophilus*. *Netherlands Milk and Dairy Journal* **4**, 197-208.
Pette, J.W. and Lolkema, H. (1950b). Yoghurt II. Groeifactoren voor *Sc. thermophilus*. *Netherlands Milk and Dairy Journal* **4**, 209-224.
Pettersson, H.-E. (1975). Growth of a mixed species lactic starter in a continuous "pH-stat" fermentor. *Applied Microbiology* **29**, 437-443.
Radler, F. (1975). In "Lactic acid bacteria in beverages and foods" (eds. J.G. Carr, C.V. Cutting and G.C. Whiting), 17-27. Academic Press.
Rogers, L.A. and Whittier, E.D. (1928). Limiting factors in the lactic fermentation. *Journal of Bacteriology* **16**, 211-229.
Shankar, P.A. and Davies, F.L. (1977). Amino acid and peptide utilization by *Streptococcus thermophilus* in relation to yogurt manufacture. *Journal of Applied Bacteriology* **43**, VIII.
Shankar, P.A. and Davies, F.L. (1978). Interrelationships of *Streptococcus thermophilus* and *Lactobacillus bulgaricus* in yogurt starters. *20th Int. Dairy Congr.*, vol.E, 514-515.

Tempest, D.W. (1970). The continuous cultivation of micro-organisms
I. Theory of the chemostat. *Methods in Microbiology* **2**, 259-276.
Veringa, H.A., Galesloot, Th.E. and Davelaar, H. (1968). Symbiosis
in yogurt (II). Isolation and identification of a growth factor
for *Lactobacillus bulgaricus* produced by *Streptococcus thermophilus*. *Netherlands Milk and Dairy Journal* **22**, pp. 114-120.
Walstra, P. and Van der Haven, M.C. (1979). Melkkunde. Een inleiding in de samenstelling, structuur en eigenschappen van melk.
L.H.-Wageningen p.8.
Wilkinson, T.G., Topiwala, H.H. and Hamer, C. (1974). Interaction
in a mixed bacterial population growing on methane in continuous
culture. *Biotechnology and Bioengineering* **16**, 41-59.
Wood, H.G. and Stjernholm, R.L. (1962). Assimilation of carbon
dioxide by heterotrophic organisms. The Bacteria. (eds.
I.C. Gunsalus and R.Y. Stanier), Vol. III, pp.41-117. New York-London: Academic Press.

Chapter 6
YEAST–BACTERIUM INTERACTIONS
IN THE BREWING INDUSTRY

F.H. WHITE and E. KIDNEY

Bass Limited, Burton-on-Trent, Staffs. UK

1. Introduction

Some of the more obvious factors influencing the susceptibility of beer to microbial spoilage have been itemized [Rainbow, 1981] as: (i) the level in the beer of nutrients such as carbohydrates, nitrogenous compounds, salts, vitamins, and growth factors;   (ii) the levels of the products of yeast metabolism, mainly ethanol, carbon dioxide, higher alcohols, esters and organic acids;   (iii) the pH and oxygen content of the beer;   and (iv) the concentration of hop bitter substances.

With the exception of the latter, all these factors reflect the extent and vigour of yeast growth and fermentation. Yeast activity therefore exerts a major influence on the tendency of beer to spoil. Thus, competition between yeast and bacteria for available wort nutrients has been suggested as the basis for the varying susceptibility of fermentations to spoilage by *Hafnia protea* [Strandskov and Bockelmann, 1955, 1956, 1957]. In contrast, instances in which bacterial growth was dependent on nutrients synthesized and excreted by yeast have been reported [Rose, 1954; Challinor and Rose, 1954].

In this chapter, however, we wish to describe another means by which certain yeast strains can influence the tendency of beer to microbial spoilage: by aggregating and cosedimenting with specific bacterial cells.

2. Aggregation Amongst Microorganisms

The ability of bacteria in nature to adhere to other cells and to a wide range of biological and non-biological surfaces [Harris and Mitchell, 1973], is an important factor in the aetiology of many human [Gibbons and Van Houte, 1975] and animal [Costerton *et al*., 1978] diseases, as well as in the design of certain industrial fermentation processes [Atkinson and Daoud, 1976].

Aggregation among brewing yeasts is manifested most frequently in the phenomenon of flocculence, which describes the strain-dependent ability of yeast cells to aggregate together, and either fall out of suspension or be carried to the top of the fermentation broth [Rainbow, 1966]. Aggregation between heterogeneous yeast populations has also been reported [Eddy, 1958; Hough, 1959], and the term coflocculence defined [Stewart and Garrison, 1972; Stewart et al., 1973] as referring specifically to flocculation between cells of two discrete yeast strains, each of which is non-flocculent in isolation.

To date however, the only reports referring to specific interactions between yeast and bacterial cells are those describing aggregation between lactic acid bacteria and certain strains of *Saccharomyces cerevisiae* used in saké production [Momose et al., 1969], and a recent study showing agglutination of *Escherichia coli* cells by both soluble yeast mannan and intact yeast cells [Ofek et al., 1977].

3. Occurrence of Bacterial Contamination

Our studies began with an investigation into the reasons why certain fermentations consistently contained higher levels of bacteria than others in the same brewery. Preliminary studies attempted to establish a relationship between the frequency of spoilage and other properties of the particular worts of beers. When all such attempts failed, and no inadequacies of plant cleaning could be found, an examination was made of the yeast strains used in this particular plant.

4. Yeast—Bacterium Aggregations

We especially compared the ability of the different yeast strains to aggregate and cosediment with those bacteria most frequently identified at the brewery in question. The numbers of *Lactobacillus* cells in suspension in beer in the presence of two different yeast strains are given in Table 1. Although both yeast strains had comparable flocculation characteristics, the presence of cells of *Saccharomyces cerevisiae* strain B resulted in a marked decrease in the numbers of bacterial cells in suspension, whilst the presence of the same concentration of cells of *S. cerevisiae* strain A had no such effect. A similar, albeit less dramatic, reduction in the numbers of bacteria in suspension occurred in the presence of a mixture of the two yeast strains.

A further series of experiments was designed to examine the influence of yeast strain B on the distribution of bacteria throughout the suspending medium. The results obtained are summarized in Table 2, which compares the number of *Lactobacillus* cells in beer in a test-tube in the presence and absence of *S. cerevisiae*

TABLE 1

Effect of Yeast Strain on Numbers of Bacteria in Suspension

Time (day)	Viable counts (organisms ml^{-1}) of *Lactobacillus brevis* organisms in suspension when mixed in beer with:		
	Yeast A (1.0 x 10^5 organisms ml^{-1})	Yeast B (1.0 x 10^5 organisms ml^{-1})	Yeast A+B (1.0 x 10^5 organisms ml^{-1})
Day 0	7.0 x 10^2	2.3 x 10^3	1.9 x 10^3
1	7.0 x 10^2	3.0 x 10^2	4.0 x 10^2
4	3.0 x 10^2	1.5 x 10^2	2.5 x 10^2
7	6.0 x 10^2	23	1.3 x 10^2

strain B. When no yeast cells were present, the numbers of bacteria increased uniformly throughout the beer, so that after 7 days the concentration of bacteria in a sample of beer removed from the middle of the tube was approximately the same as that in a sample removed from the base of the tube. However, in the presence of cells of yeast strain B, not only did the numbers of bacteria in suspension decrease, but the concentration of bacterial cells at the bottom of the tube increased. It was thus concluded that the *Lactobacillus* cells were interacting with the yeast cells, with the result that they were removed from suspension and concentrated in the yeast sediment at the bottom of the tube.

5. Effect of Medium Composition

We decided on the term cosedimentation to describe this interaction between yeast and bacteria, and using cells of *Saccharomyces cerevisiae* strain B and *Lactobacillus brevis* as a model system, we set out to establish to what extent cosedimentation could be influenced by changes in wort composition and pH.

As shown in Table 3, although the total numbers of viable bacteria declined as beer pH was lowered, in the absence of yeast, at all but one pH value the ratio of bacterial counts per ml of suspension to counts per ml of sediment fell within the range 1/1.3 to 1/2.0. However, in the presence of cells of yeast strain B and at any value of pH over the range 3.5 to 5.0, the concentration of bacteria in the sediment was always substantially

TABLE 2

Effect of Saccharomyces Cerevisiae *Strain B on the Distribution of* Lactobacillus Brevis *Organisms in Beer*

Time (day)	Viable counts of *L. brevis* organisms ml^{-1}							
	With yeast (1.0 x 10^7 organisms ml^{-1})				Without yeast			
	in suspension	in sediment	Ratio	$\frac{\text{suspension}}{\text{sediment}}$ (organisms ml^{-1})	in suspension	in sediment	Ratio	$\frac{\text{suspension}}{\text{sediment}}$ (organisms ml^{-1})
0	3.4 x 10^3	2.7 x 10^3	1/0.8		7.6 x 10^2	1.1 x 10^3	1/1.5	
7	1.5 x 10^2	2.7 x 10^4	1/180		7.6 x 10^3	1.5 x 10^4	1/2.0	

TABLE 3

Effect of pH on Distribution of Lactobacillus Brevis Organisms in Beer

Viable counts of *L. brevis* (organisms ml^{-1}) in beer after 7 days

Beer pH	With yeast strain B (1.0 x 10^7 organisms ml^{-1})			Without yeast		
	in suspension	in sediment	Ratio $\frac{\text{suspension}}{\text{sediment}}$ (organisms ml^{-1})	in suspension	in sediment	Ratio $\frac{\text{suspension}}{\text{sediment}}$ (organisms ml^{-1})
3.5	5	3.7 x 10^2	1/74	12	16	1/1.3
4.0	3.1 x 10^2	3.3 x 10^3	1/11	1.5 x 10^4	3.0 x 10^4	1/2
4.5	1.6 x 10^5	2.7 x 10^6	1/17	6.0 x 10^6	8.0 x 10^6	1/1.3
5.0	5.2 x 10^5	4.3 x 10^6	1/8	1.4 x 10^6	7.0 x 10^6	1/5

Starting concentration of *L. brevis*: 3 x 10^3 organisms ml^{-1}

greater than the concentration in suspension. It appeared that at least within the range examined, pH did not exert a significant influence on the cosedimentation between these two cell types.

Cosedimentation with *L. brevis* cells was also observed during the fermentation of all-malt wort by *S. cerevisiae* strain B, the decrease in the number of viable bacteria in suspension being of the same order as the decrease in fully-fermented beer of the same period. A study of the influence of the various wort fermentable sugars showed that cosedimentation between yeast strain B and *L. brevis* cells occurred to a comparable extent in all-malt wort in which 40% of the extract was derived from maltose (Table 4). However, cosedimentation occurred to a notably lesser extent when either glucose or sucrose were used to supplement all-malt wort.

TABLE 4

Influence of Wort Sugars on Cosedimentation of Saccharomyces Cerevisiae *Strain B with* Lactobacillus Brevis

Wort Composition	Viable counts of *L. brevis* (organisms ml^{-1}) in suspension			
	With yeast (1.0×10^7 organisms ml^{-1})		Without yeast	
	Day 1	Day 8	Day 1	Day 8
All-malt	4.6×10^3	5.0×10^2	1.8×10^3	5.1×10^6
40% maltose	5.7×10^3	1.9×10^2	2.9×10^3	5.1×10^6
40% glucose	3.6×10^3	1.2×10^3	2.1×10^3	4.7×10^6
40% sucrose	2.9×10^3	8.8×10^2	2.8×10^3	2.3×10^6

6. Turbidometric Assay

In order to look for other examples of cosedimentation between strains of brewing yeast and species of the commonly occurring beer spoilage bacteria, a rapid but sufficiently sensitive turbidometric assay procedure was developed for screening various combinations of yeast and bacteria.

The suspending medium used in these studies consisted of 50 mM acetate buffer, pH 4.6, containing 0.1% (w/v) calcium chloride [Mills, 1964]. The turbidities of yeast and bacterial suspensions were measured over a 15 min period, both separately and after mixing, using an EEL Colorimeter, fitted with a Chance OB4 filter.

With most combinations of cells traces such as those shown in Figs 1 and 2 were obtained, demonstrating that the turbidities of mixed suspensions were equal to the sum of the turbidities of the individual yeast and bacterial suspensions. In such instances there was therefore nothing to indicate an interaction between the respective cell types.

However, with certain combinations of yeast and bacteria the turbidity of the mixed suspension was found to be below not only the sum of the individual turbidities, but also the turbidity of the yeast suspension alone (Figs 3 and 4). Such instances were assumed to reflect cosedimentation between the respective yeast and bacterial cells. In a few cases cosedimentation was so strong that the turbidity of the mixed suspensions was less than either of the individual cell suspensions (Fig. 5).

In order visually to assess the extent of cosedimentation between yeast and bacteria, it was necessary to adjust the concentrations of the individual cell suspensions so that they each had turbidities on the same 0 to 10 scale. Since this necessitated the use of quite concentrated bacterial suspensions, it meant that the ratio of yeast cells/bacteria in the resulting mixed suspensions was reduced from the standard "in excess of 100/1" maintained throughout previous studies to values of between 1/1 and 1/10 yeast/bacteria. However, in all instances, cosedimentation occurring at these ratios was, on inspection, also found to occur at the higher ratios of yeast cells to bacteria, such as would be found in a brewery.

The results of a limited survey of five brewing strains of *Saccharomyces cerevisiae* tested against five species of commonly occurring brewery bacteria are shown in Table 5.

Cosedimentation appeared to be a far more widespread phenomenon than at first realized and, when it occurred, was apparently specific for the individual yeast and bacteria. *S. cerevisiae* strain B was exceptional among the yeast strains tested in cosedimenting strongly with four out of five bacteria, including both Gram-positive and Gram-negative species. Of the bacteria, *Hafnia protea* appeared to possess the broadest spectrum of activity.

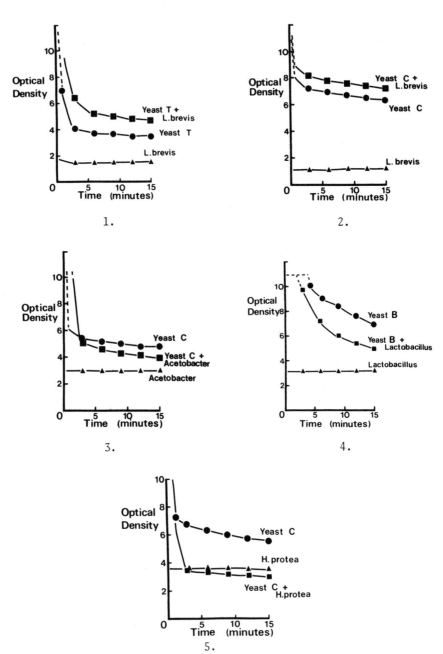

Figs. 1-5. Turbidometric measurement of yeast/bacteria interactions

TABLE 5

Incidence of Cosedimentation between Brewing Strains
of Saccharomyces Cerevisiae and Beer Spoilage Bacteria

Yeast strain	Hafnia protea	Lactobacillus brevis	Lactobacillus sp.	Pediococcus sp.	Acetobacter sp.
S. cerevisiae, strain B	+++	+++	+++	-	++
S. cerevisiae, strain C	+++	-	-	+	++
S. cerevisiae, strain T	+++	-	-	+	-
S. cerevisiae, strain 2	++	++	-	-	-
S. cerevisiae, strain 3	+	+	+	+	+

- no interaction, + slight interaction, ++ strong interaction, +++ very strong interaction

7. Role of Calcium

The turbidometric assay procedure was used in subsequent studies to assess the influence on cosedimentation of divalent cations, especially calcium, in the suspending medium, and changes to various functional groups on the yeast cell wall. *Saccharomyces cerevisiae* strain C and *Hafnia protea* were chosen as the test system for this study.

As shown in Fig. 6, thorough washing of *S. cerevisiae*, strain C after growth in brewers' wort, followed by resuspension in calcium-free acetate buffer, substantially reduced its ability to cosediment *H. protea*. However, the addition of 0.1% (w/v) calcium chloride to the suspending buffer led to immediate restoration of the cells cosedimenting capabilities.

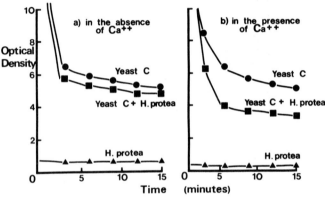

Fig. 6. Cosedimentation of *Saccharomyces cerevisiae* strain C with *Hafnia protea*.

Similar results were obtained after treating the yeast cells with EDTA prior to assaying. Further confirmation of the role of calcium in the interaction between these two cell types was obtained from studies in which both yeast and bacteria were grown in defined medium. No cosedimentation occurred between such cells in the absence of calcium. The addition of Mg^{2+}, Na^+, or K^+ ions to cell suspensions in acetate buffer resulted in no observable changes in cosedimentation characteristics.

8. Cell Wall Protein Composition

When cells of *Saccharomyces cerevisiae* strain C were treated with pepsin, not only was their ability to flocculate destroyed, so that the optical density of the yeast suspension remained constant over the 15 min assay period, but their ability to cosediment *Hafnia* cells was also abolished (Fig. 7). Thus, although the turbidity of a mixed suspension of control (untreated) yeast cells and *H.protea* was significantly less than the turbidity of the yeast suspension alone, the turbidity of a mixed suspension of pepsin-treated yeast cells and bacteria was equal to the sum of the individual turbidities.

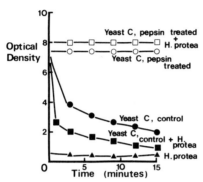

Fig. 7. Effect of pepsin on flocculation and cosedimentation of *Saccharomyces cerevisiae* strain C with *Hafnia protea*.

Treatment of *S.cerevisiae* strain C with trypsin was without any measurable effect on its ability to cosediment *H.protea*. However, the effect of chymotrypsin treatment was the same as that of pepsin — a complete inhibition of both flocculation and cosedimentation.

The results with pepsin and chymotrypsin therefore indicated an essential role for yeast cell wall protein components in cosedimentation as well as in flocculation. In addition, the known preference of both these proteolytic enzymes for peptide linkages involving aromatic amino acids [Mahler and Cordes, 1971], suggested a requirement for these specific amino acids in the cosedimentation reaction.

Treatment of *S.cerevisiae*, strain C with other reagents known to restrict or inhibit flocculation, was mostly without effect on cosedimentation. Thus, as summarized in Table 6, α-amylase action and prolonged oxygenation of yeast cells in buffer, both of which treatments inhibited flocculation [thus confirming the findings of Stewart et al., 1973], had no effect on cosedimentation. Similarly either esterification of surface carboxyl and phosphate groups, deamination of free amino groups, or reduction of disulphide bonds, although inhibiting flocculation [Nishihara et al., 1977], were without effect on cosedimentation. Only acetylation, using N-acetylimidazole and urea, completely inhibited both flocculation and cosedimentation (Fig. 8).

It was noteworthy that urea was required in the

TABLE 6

Effect of Chemical Modification of the Yeast Cell Wall on Flocculation and Cosedimentation

Treatment	Effect on flocculation:	Effect on cosedimentation:
E.D.T.A.	Restricted	Restricted
Pepsin	Inhibited	Inhibited
α-Amylase	Inhibited	No effect
Starvation	Inhibited	No effect
Esterification	Inhibited	No effect
Deamination	Inhibited	No effect
Disulphide bond reduction	Restricted	No effect
Acetylation	Inhibited	Inhibited

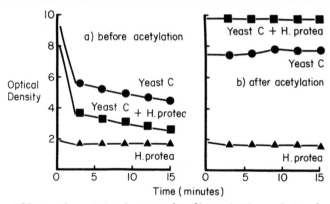

Fig. 8. Effect of acetylation on the flocculation of *Saccharomyces cerevisiae* strain C and its cosedimentation with *Hafnia protea*.

reaction mixture in order to obtain a complete inhibition of cosedimentation. Treatment with N-acetylimidazole alone, although inhibiting flocculation, resulted in only a partial inhibition of cosedimentation. Again therefore, the proteinaceous nature of the reactive site on the yeast cell wall was indicated. Moreover, the known specificity of N-acetylimidazole activity towards the hydroxyl groups on the phenolic side-chain of tyrosine residues [Mahler and Cordes, 1971], in addition to further implicating aromatic amino acids, suggested a specific role for the tyrosine residues in yeast cell wall proteins.

Deacetylation of acetylated cells resulted in complete restoration of both flocculation and cosedimentation, demonstrating that the N-acetylimidazole was actually blocking reactive sites on the yeast cell surface, and not causing their loss into the suspending medium.

The nature of the interaction between cells of *S.cerevisiae* strain C and *H.protea* was therefore different to the aggregation phenomena previously observed [Momose et al., 1969] between *S.cerevisiae* species and lactobacilli, when electrophoretic studies showed that reactive organisms had isoelectric points between pH 3.5 and 4.0, and that aggregation only occurred when the pH of the suspending buffer was such that the net surface charge on the yeast and bacterial cells was opposite [Momose et al., 1969]. Electrophoretic studies by us on a number of cosedimenting yeast and bacteria revealed that all the cell types had the same (negative) charge between pH 3.6 and 5.6, thus confirming earlier studies (Table 3) that changing the pH of the suspending medium over this range had no effect on cosedimentation.

No significant effect of temperature on cosedimentation between *S.cerevisiae* strain C and *H.protea* could be observed. However, cosedimentation was influenced by the age of the yeast cell. Thus, as shown in Fig. 9, cells harvested in late log phase, after 18 h growth, exhibited the strongest cosedimentation, whilst yeast harvested from 7 days old cultures, reacted only poorly. In contrast, when the age of the bacterial cells was varied, strongest cosedimentation was observed with stationary phase cells.

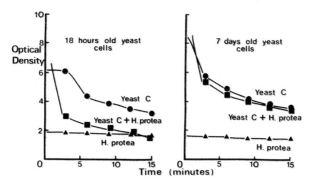

Fig. 9. Cosedimentation of *Saccharomyces cerevisiae* strain C with *Hafnia protea*.

Conclusion

In conclusion, it may be worthwhile to consider some of the implications of these findings to practical brewing. The immediate effect of any change in the distribution of bacteria due to cosedimentation, is to reduce the numbers of bacteria carried over with the beer into the conditioning tank. However, this short term benefit is more than offset by the tendency of cosedimentation to concentrate and perpetuate bacteria in the yeast crop.

Although the influence of yeast wall proteins and calcium have been demonstrated, cosedimentation is not significantly influenced by any practicable changes in wort composition or the manner in which fermentations are conducted. Rather it is suggested that any combative procedures be based on preventing access of those bacteria cosedimenting with the particular yeast strain in use. Equally, more frequent introduction of pure cultures and/or recourse to acid washing may well be advisable in cases where this effect is widespread.

Acknowledgements

The authors wish to thank the Directors of Bass Limited for permission to publish these results.

References

Atkinson, B. and Daoud, I.S. (1976). Microbial flocs and flocculation in fermenting process engineering. In "Advances in Biochemical Engineering", Vol 4 (eds. T.K. Ghose, A. Fiechter and N. Blakeborough), pp.41-124. Berlin and New York: Springer Verlag.

Challinor, S.W. and Rose, A.H. (1954). Interrelationships between a yeast and a bacterium when growing together in a defined medium. *Nature, London* **174**, 877-878.

Costerton, J.W., Geesey, G.G. and Cheng, K.-J. (1978). How bacteria stick. *Scientific American* **238**, 86-95.

Eddy, A.A. (1958). Composite nature of the flocculation process of top and bottom strains of saccharomyces. *Journal of the Institute of Brewing* **64**, 143-151.

Gibbons, R.J. and van Houte, J. (1975). Bacterial adherance in oral microbial ecology. *Annual Review of Microbiology* **29**, 19-44.

Harris, R.H. and Mitchell, R. (1973). The role of polymers in microbial aggregation. *Annual Review of Microbiology* **27**, 27-50.

Hough, J.S. (1959). Flocculation characteristics of strains present in some typical British pitching yeast. *Journal of the Institute of Brewing* **65**, 479-482.

Mahler, H.R. and Cordes, E.H. (1971). In "Biological Chemistry", pp.117-341, 2nd edn. London and New York: Harper and Row.

Mills, P.J. (1964). The effect of nitrogenous substances on the time of flocculation of *Saccharomyces cerevisiae*. *Journal of General Microbiology* **35**, 53-60.

Momose, H., Iwano, K. and Tonoike, R. (1969). Studies on the aggregation of yeast caused by Lactobacilli 4: Force responsible for aggregation. *Journal of General and Applied Microbiology* **15**, 19-26.

Nishihara, H., Toraya, T. and Fukui, S. (1977). Effect of chemical modification of cell surface components of a brewer's yeast on

the floc forming ability. *Archives of Microbiology* **115**, 19-23.
Ofek, I., Mirelman, D. and Sharon, N. (1977). Adherence of *Escherichia coli* to human mucosal cells mediated by mannose receptors. *Nature, London* **265**, 623-625.
Rainbow, C. (1966). Flocculation of brewer's yeast. *Process Biochemistry* **2**, 489-492.
Rainbow, C. (1981). Beer spoilage microorganisms. In "Brewing Science" (ed. J.R.A. Pollock) London and New York: Academic Press.
Rose, A.H. (1954). Ph.D. Thesis, Birmingham University.
Stewart, G.G. and Garrison, I.F. (1972). Some observations on coflocculation in *S.cerevisiae*. *Proceedings of the American Society of Brewing Chemists*, 118-131.
Stewart, G.G., Russell, I. and Garrison, I.F. (1973). Further studies on flocculation and coflocculation in brewer's yeast strain. *Proceedings of the American Society of Brewing Chemists*, 100-106.
Strandskov, F.B. and Bockelman, J.B. (1955). Nutritional requirements of brewing microorganisms: 1. The nutritional requirements of *Flavobacterium proteus*. *Proceedings of the American Society of Brewing Chemists*, 36-42.
Strandskov, F.B. and Bockelman, J.B. (1956). Effect of brewer's yeast strain on *Flavobacterium proteus* contaminants of brewery fermentations. *Wallerstein Laboratories Communications*, 313-319.
Strandskov, F.B. and Bockelman, J.B. (1957). The effect of yeast strain on bacterial growth in brewery fermentation. *Proceedings of the American Society of Brewing Chemists*, 94-97.

Chapter 7
THE YEAST/LACTOBACILLUS INTERACTION; A STUDY IN STABILITY

B.J.B. WOOD

*Department of Applied Microbiology,
University of Strathclyde, 204 George Street, Glasgow, UK*

1. Introduction

In this paper I wish to show that associations between yeasts and lactic acid bacteria are very common in a wide variety of traditional food and beverage fermentations. In some cases, for example sour-dough levens, they are also remarkably stable, seemingly persisting unchanged for many years if appropriate provision is made for them. A further remarkable, and technologically important characteristic of these associations, is their ability to achieve dominance over the organisms naturally present in the raw materials used in the fermentations. Clearly, the alcohol produced by the yeast, the acids produced by the bacteria and the anaerobiosis induced by the fermentation will all contribute to the suppression of some at least of the other microbes present in the system, but it seems hard to believe that this is always an adequate explanation of the success of these associations. Table 1 lists some typical fermentations in which yeasts and lactic acid bacteria are known or believed to participate jointly. I propose to discuss each group in fairly general terms, drawing attention to particularly significant aspects of the case under examination.

2. Beverages

Saké, the traditional rice "wine" of Japan is the product of a two-stage fermentation in which polished rice is cooked, inoculated with a strain of *Aspergillus oryzae* which produces high yields of amylase, incubated for about 50 h until covered with mould mycelium, then mixed with water, when the second stage fermentation occurs [Wood, 1977]. In this second fermentation, conditions are so arranged that a lactic fermentation is

TABLE 1

Representative Yeast/Lactic Acid Bacteria Fermentations

Name	Raw materials	Country of origin
Beverages		
Saké	Polished rice	Japan
Lambic beer	Malted barley	Belgium
Geuze beer	Malted barley	Belgium
Stock beer	Malted barley	England (obsolete)
Ginger "beer"	Ground ginger, sugar	United Kingdom
Whisky	Malted barley	Scotland
Sourmash Bourbon	Cereals	USA
Kvass	Toasted rye bread	Russia, E. Europe
Kaffir beer, etc.	Various grains	Africa
Milk Based		
Koumiss	Mare's milk	Russia
Kefir	Cow's milk	Russia, E. Europe
Foods		
Various cassava fermentations	Cassava	Africa, S. America
Sour-dough breads	Wheat and rye	World wide
Parisian barm bread leven	Wheat	Scotland
Flavourings		
Soy sauce	Soy beans, sometimes + wheat	China, Japan, S.E. Asia
Miso and other fermented soy pastes	Soy beans, sometimes + barley or rice	China, Japan, S.E. Asia
Animal feed		
	Feedlot waste + corn	USA (novel process)

developed which is sufficient to give a pH appropriate for yeast growth and also for the desired flavour of the final product. Too free a production of acid would prevent growth of the yeast and must therefore be avoided. The conditions favour the continued activity of the mould amylases, and so a steady supply of fermentable sugar is provided and its rapid fermentation to acid and (mainly) alcohol ensures that the sugar concentration is always kept quite low, so permitting the fermentation to attain a high concentration of alcohol.

In the Belgian Lambic and Geuze beers and in the old stock beer of England the participation of lactic acid bacteria gives a rather sour beer by most standards [Lloyd Hind, 1950]. The Belgian beers take a long time to ferment (up to two years) and involve the participation of several organisms including *Saccharomyces* and *Brettanomyces* yeasts and species of lactic acid bacteria.

The part played by lactic acid bacteria in sorghum beer and similar African beverages is similar to their part in the Belgian beers, giving a sharpness to the flavour of the beverage which is said to be particularly refreshing in the heat of Africa. Modern technology results in a much more controlled fermentation, resulting in a product of consistently high quality [Novellie, 1980]

The production of kvass has much in common with the sour-dough bread fermentations so common in the regions whence it comes, principally Russia and Eastern Europe. Indeed it seems that a portion of bread leven may be used as a starter for a fresh kvass fermentation. The basic raw materials for the fermentation are fresh water and bread. The bread is toasted or rusked, crumbled and mixed with the water, then this mixture is used to replace liquid drawn off from a vessel of fermenting brew. Traditionally this liquid is drunk at once, and is said to have a sharp, refreshing flavour and mild carbonation. Recently there has been an interest in Poland in producing a clarified, pasteurized, carbonated version of this beverage, bottled for sale in supermarkets and shops in competition with imported American soft drinks [Z. Wlodarczyk, pers. comm.]. This drink has a very pleasant taste, having a mild, sharp, nutty flavour with a hint of the rye bread used in its manufacture. The carbonate is sufficient to give it a sparkle on the palate without excessive "gassiness", and the liquid is an attractive golden brown colour. The drink is not obviously alcoholic and its low sweetness makes it much more acceptable than most soft drinks. It will be realized that in the traditional method of fermentation there would be a high chance of contamination with undesirable microorganisms, but in practice it seems that the fermentation rarely gives any trouble provided that it is supplied with fresh materials at reasonably regular intervals. There are many traditions asserting that kvass, in common with kefir and koumiss, is a beverage conferring benefit upon the user in a variety of ways. In particular it is reputed to benefit the digestive tract and it has even been suggested that it can afford some protection against cancer. It is not at all clear how far any of these claims would survive the scrutiny of Western scientific medicine, however.

I am not aware of any modern examination of the microbiology of ginger beer production using the traditional ginger beer "plant", but there are said to be very old reports indicating that this is a yeast/lactic fermentation, and consideration of the process involved (which has similarities with the production of kvass and of sourdough levens) suggests that this is a plausible hypothesis. The process is a little difficult to describe although it is very easy to carry out. The starter culture, called a ginger beer "plant", is a rather slimy, muddy-looking lump of material. This is placed in two cups of water and "fed" with a little sugar and ground ginger (usually 1 teaspoon of each) per day for about two weeks. A lively fermentation takes place. At the end of the two weeks the liquid is strained off, mixed with about a gallon of water containing 3½ cups of sugar and the juice of three lemons, and then bottled; after about 10 days a refreshing, sparkling, slightly alcoholic liquid stands over a slight, slimy deposit. Meanwhile the "plant" is divided into two equal portions and treated as before. Now ground ginger, like most commercial spices, carries an appreciable load of microbes, yet this fermentation can be kept going without the slightest difficulty using the method described above. The viscous nature of the "plant" is reminiscent of kefir and koumiss, where the characteristic "grains" are bound together by extracellular polysaccharide elaborated by the microorganisms. In short then, this fermentation seems to have characteristics of microbial stability and the production of extracellular polysaccharides which are found in typical yeast/lactic fermentations. Rather less typical is the somewhat limited nutrition afforded by the medium, in contrast with the richer and more nutritionally varied fare offered by some of the other brews. Microbiological and biochemical studies of this fermentation would be most interesting, and the sequence of microflora which developed during an attempt to start a "plant" *de novo* from sugar, water and ground ginger could be illuminating.

Finally, whisky; the name "sourmash bourbon" makes clear the importance of controlled souring in this traditional American drink, but the desirability of lactic acid bacteria in Scotch whisky fermentation is less definite. Excessive development of these organisms results in souring of the fermenting liquid ("mash") to the extent that the yeasts are prevented from fermenting, so producing a "stuck mash" a very undesirable event in an industry which operates on very narrow profit margins. However, given the conditions under which most whisky fermentations are carried out, it is practically impossible to exclude all lactic acid bacteria and a clear ecological sequence of these organisms has been demonstrated [Bryan-Jones, 1975; Dolan, 1979]. While some distillers probably regard any lactic acid bacteria with suspicion, there is a fairly wide belief that in moderation they are beneficial, with the products of their fermentation having a contribution to the generics which give the whisky its desired flavour and aroma. The extent of interaction (if any) between the yeast and lactic acid bacteria in the fermentation vessel seems to be very uncertain at present.

3. Foods

Information on African cassava fermentations is steadily appearing in the West as a result of the general increase in traditional fermentations aided by conferences such as GIAM V and the Fermented Foods Sessions at the recent VIth International Fermentation Symposium, in London, Ontario. According to Sẽnora Olga Cardenas [pers. comm.], similar fermentations are practiced by the rural populations of Colombia. In general the fermentations are carried out in primitive conditions such that after a short initial period of aerobic fermentation, anaerobic conditions prevail and numerous microorganisms contribute acids and flavour volatiles as a result of their fermentative activities. It seems probable that yeasts and lactic acid bacteria soon come to dominate these fermentations, resulting in a sort of dough which is cooked in various ways [Akinrele et al., 1963].

Bread levens involving lactic acid bacteria as well as yeasts are important agents in traditional bread-making throughout the world, and their use probably pre-dates the start of recorded history. Until very recently the baker's only choice was between these levens and the yeast remaining from the production of alcoholic beverages. The advent of the now-familiar pressed and dried bakers yeasts of commerce almost totally displaced the traditional levens in British and United States baking. Happily, these arts (with the exception of the Scottish process which used a liquid leven called a Parisian Barm — Bennion, 1954) were never totally lost and such speciality breads as sourdough rye bread, made mainly for the immigrant Jewish communities in the United Kingdom, are now finding a wider market as a result of the increased interest in "natural" foods. The sourdough breads of San Fransisco, California have always enjoyed a good market there, and are now available all across the USA. I was also interested to see sourdough rye bread on sale in the supermarkets of Singapore.

The work of Kline and Sugihara [1971, 1973], Kline et al. [1970] and Sugihara et al. [1970, 1971] has shown that the levens used in California are a very stable association between particular yeasts (*Saccharomyces inusitatus* and *Torulopsis holmii* have been reported) and a particular bacterium (*Lactobacillus sanfrancisco*). Similar organisms are found in sourdough levens from a Polish-Jewish Bakery in Glasgow [Wood et al., 1975] and bakeries in Poland [B.J.B. Wood, Z. Wlodarczyk, unpub. obsv.].

The yeast component of this very stable association is unable to assimilate maltose (the principal sugar in flour) but uses glucose readily. It is also very resistant to the fungal antibiotic cycloheximide. The lactobacillus utilizes maltose by the maltose phosphorylase pathway first reported in lactobacilli by Wood and Rainbow [1961] who were working with brewery isolates. This metabolic route yields glucose as a by-product.
Z. Wlodarczyk [unpub. work] has shown that the yeast utilizes this glucose and in turn it releases nutrients to the bacterium. In this way, a balance is kept between the two components of the microbial system. How this association developed is not clear at present. If a fresh sourdough is started with flour and water, a population of yeasts and lactic acid bacteria soon predominates, but in our hands these are not the specific organisms found in long-established levens.

In technical practice the maintenance of a leven is a fairly simple matter. At regular intervals, generally between 8 and 24 h, the leven is divided. One part goes to inoculate the dough for a batch of bread. The other part is also mixed with dough and held in a trough or other suitable container until required; this is the new leven. These processes are of course carried out under clean and hygienic conditions, but the system is open to the air and is far from aseptic (as the microbiologist would understand this term). The Pioneers of Alaska, the Yukon and the Far West of the United States of America were known as "Sourdoughs" from their practice of carrying and using a sour dough starter for their bread and pancakes. Clearly their lifestyle did not permit the achievement of high standards of hygiene, yet their sourdough levens generally remained wholesome and are the ancestors of the levens used in San Fransisco to this day, so demonstrating the remarkable stability and robustness of this microbial association.

German baking practice also uses sour levens, and

their microbiology has been subjected to detailed scrutiny by a group of workers at the Bundesforschungsanstalt für Getreide-und Kartoffelverarbeitung, Detmold [Spicher, 1959, 1974; Spicher and Stephan, 1964, 1966; Spicher and Schröder, 1978, 1979a,b, 1980; Spicher et al., 1979; Spicher et al., 1980]. They find a much more diverse array of bacteria, including representatives of the subgroups Thermobacterium, Streptobacterium and Betabacterium. In different levens different bacteria predominated. Similarly, the yeasts exhibited great diversity, including (in isolates from the levens of only two bakers) *Candida krusei* (27 strains), *Saccharomyces cerevisiae* (11 strains), *Pichia saitoi* (2 strains) and *Torulopsis holmii* (4 strains). Of these, only the isolates assigned to *S. cerevisiae* were able to utilize maltose. These interesting differences between the Detmold group's results and those described earlier are worthy of more detailed examination.

4. Flavourings

The microbiology and biochemistry of soy sauce and miso have been studied for many years in the Orient, but have only recently been subjected to detailed scrutiny by Occidental scientists [Yong and Wood, 1974, 1976, 1977a,b; Wood and Yong, 1975; Wood, 1977; Abiose, 1980; Abiose et al., 1981]. Here the relationship is sequential rather than associative, with the lactic acid bacteria (*Pediococcus* sp. and *Lactobacillus* sp.) appearing first and lowering the pH of the salty moromi (salt mash) second stage of this complex fermentation, to the point where the yeast (generally *Saccharomyces rouxii*) can develop. There seems to be little interaction between the yeast and bacterial components of this fermentation. However, we do not believe that the role of the bacteria is entirely restricted to the provision of lactic acid. When a moromi is acidified with lactic acid then subjected to fermentation by yeast alone, the resulting product is of good quality but is perceptibly inferior to that made by a yeast/lactic acid bacteria fermentation. We do not yet know the nature of this difference, and the task of chemical analysis of these complex mixtures to the level of discrimination involved in organaleptic assessments is probably beyond the capability of currently existing instruments and analytical methods.

For the purposes of the present paper, it must suffice to point out that once again, yeasts and lactic acid bacteria are operating under traditionally rather uncontrolled conditions in such a way as to convert a rather bland material (soy beans, wheat flour, barley or rice, *Aspergillus oryzae* mycelium and salt) into a piquant and strongly flavoured food ingredient.

5. Animal Feed

This rather modern application of the yeast/lactic acid bacteria association is perhaps the most remarkable example of its capacity to overwhelm other organisms to the benefit of the people operating the technology. The study developed as part of a U.S. Department of Agriculture, Northern Regional Research Laborototies, Peoria, programme on the use of "feedlot waste", essentially the faeces and urine resulting from the intensive rearing of cattle and pigs [Rhodes and Hrubant, 1972; Hrubant et al., 1972; Sloneker et al., 1973; Hrubant, 1973, 1975; Weiner and Rhodes, 1974; Rhodes and Orton, 1975; Weiner, 1977a,b; Hrubant et al., 1978; Hrubant and Detroy, 1980].

The process is simple to operate, although the microbiology is complex. Basically, the feedlot waste is homogenized, screened to remove fibrous material, then mixed with coarsely cracked corn (maize) in a proportion of two parts grain to one part feedlot waste liquor (FLWL) by weight. A rotating fermentor is used to carry out the fermentation; it is of simple design and no special provision for aeration is made; conversely no provision for anaerobiosis is made either, and in practice it seems probable that microaerophyllic conditions prevail. The indigenous enteric bacteria present in the mixture (such as coliforms and faecal streptococci) disappear rapidly from the mixture, as do the betabacteria predominant among the lactics of the original wastes and also the dominant natural yeast (*Trichosporon cutaneum*). During the first 24 h fermentation a single heterolactic species becomes dominant (95% of the lactobacilli and more than 90% of the total microflora). Thereafter a consortium of three lactobacilli (one each from the subdivisions Streptobacterium, Thermobacterium and Betabacterium) and three yeasts (two *Candida* sp. and one *Pichia* sp.) develops. When swine waste was used, a similar pattern of microbial events took place, with drastic reduction in the number of coliform organisms. Yeasts, however, were only minor numerical contributors to fermentations involving swine wastes.

The technological importance of this class of fermentations resides in the ability to economize on nitrogen, converting the nitrogenous compounds normally lost in the faeces and urine to a form acceptable to the animals as part of their diet, while at the same time getting rid of a potentially highly objectionable product of intensive animal rearing, converting a foetid material into something with an odour likened to that of silage. Since the fermentation relies on natural microflora, it is suitable for operation alongside other farm operations by a workforce without skills or training in microbiology. The full spectrum of biochemical abilities deployed by organisms participating in this type of fermentation has yet to be explored; for example, changes in vitamins and other growth factors have yet to be fully catalogued. Already, however, useful discoveries are being made; the isolation of two new species of *Lactobacillus*, both of which release extracellular amylase, is a particularly interesting observation.

In sum, this type of fermentation is going to make important contributions both on a practical level and to our knowledge of the biochemical capabilities of the microbes participating in it.

6. Milk-based Foods

Kosikowski [1977] provides interesting descriptions of the two principal fermented milk-based foods which involve both yeasts and lactic acid bacteria, namely *Kefir* and *Koumiss*. In introducing the subject, he claims that the microbes involved in these foods and in yogurt release antibiotics into the medium, and urges a re-examination of the dietary importance of fermented milk foods (preferably with the cream removed).

The peoples of the Soviet Union apparently consume about 4.5 kg of kefir per head per year and similar rates may apply in some countries of Eastern Europe. As supplied, its sharp, rather cheese-like smell and taste and viscous, slimy appearance and feel are rather disconcerting to a Westerner, but once the taste for it has been acquired it is a pleasing food, good by itself but (in my opinion) much better with fresh fruit. A unique feature of kefir is that the organisms form firm granules ranging up from the size of a grain of rice. These granules consist of masses of the organisms bound together by a polysaccharide called kefiran which is composed of glucose and galactose residues. To make a batch of kefir, grains are placed into milk (which should have been pasteurized or boiled and then cooled to about 22°C). After incubation overnight at the same temperature, the milk has fermented to yield a smooth curd which, although quite acidic (up to 1% as lactic acid) shows no tendency to separate into curds and whey.

The kefir is passed through a sieve to recover the grains, which can be stored for long periods in sterilized 0.85% sodium chloride solution, or washed in water and lyophilized, or used at once to ferment a further batch of milk. The grains increase in size and number as they are repeatedly used to ferment batches of milk, and they can eventually be sub-divided into two portions each as large as the original sample of grains. Many organisms are also present in the kefir after filtering, and this can be used to ferment a large batch of milk, or stored in order to ferment further and develop more acid and a stronger flavour. As I understand it, not only is kefir regarded as being of considerable therapeutic value, but also its clinical uses vary depending upon the number of

days for which fermentation is permitted to continue.
Kosikowski [1977] lists the microbes of kefir as comprising two yeasts, *Saccharomyces kefir* and *Torulopsis kefir* and the bacteria *Lactobacillus caucasicus*, *Leuconostoc* spp. and "lactic acid streptococci". "Yeasts represent 5 to 10% of the microbial population". He notes that the grains may be "covered with a white mould, *Geotrichum candidum*", although this "apparently does not detract much from quality". He states that a typical kefir will contain about 0.8% lactic acid, 1% ethanol and sufficient carbon dioxide to cause the kefir to "foam and fizz like beer", although in trials with Polish kefir grains I have not observed such vigorous evolution of gas. Diacetyl, acetaldehyde and complex alcohols are apparently the chief flavour components in kefir.

In experiments carried out in collaboration with visiting Polish colleagues, I have made kefir of good organoleptic quality from pasteurized Scottish milk, but for some reason the kefir grains did not increase in quantity to the expected extent; indeed very little increase in quantity occurred even though the grains were transferred to fresh milk every day for several months. The explanation for this eludes us at present. It is unfortunate that this is so, since I believe that kefir could command a modest market through the "whole-food" trade.

Koumiss is classically made from mares' milk, although Kosikowski [1977] reports that a satisfactory "koumiss" is now made from skimmed cows' milk. Traditionally, the fermentation is conducted by inoculating pasteurized milk cooled to 28 to 30°C with a portion of a mature batch of koumiss. As with kefir, the koumiss shows no tendency to separate into curds and whey. The organisms involved are *Lactobacillus bulgaricus* and "Torula yeast" according to Kosikowski, who stresses the pleasing and refreshing flavour of the fermented product, which contains 0.7 to 1.8% lactic acid and 1.0 to 2.5% ethanol (w/v), depending on the type of koumiss. He also stresses the therapeutic value attached to the product in the USSR, especially in the treatment of pulmonary tuberculosis, where the recommended daily dose of 1.5 U.S. quarts (1.42 litres) induces in the patient "some excitement followed by a slight sense of intoxication and somnolence"; the intoxication apparently never causes a hangover.

I have no experience of this product, but if a version could be made in the West from cows' milk, it should find a market in the healthfood and wholefood trades.

7. Antibiotic Activities of Lactic Acid Bacteria

I am grateful to Mrs Napa Lotung of Kessetsart University, Bangkok, Thailand, for drawing my attention to the extensive documentation which now exists pertaining to the antibiotic and therapeutic activities of lactic acid bacteria. Typical reports are those of Aritaki and

Ishikawa [1962] on Japanese pediatric applications of *Lactobacillus acidophilus* milk; Hamdan and Mikolajck [1974] on Acidolin, an antibiotic produced by the same bacterium; and Shahani et al. [1977] on the natural antibiotic activity of *L. acidophilus* and *Lactobacillus bulgaricus*. The fact that these familiar organisms produce antibiotics, strongly suggests that the bacteria and yeasts which come to predominate in such exotic environments as cracked maize mixed with feedlot waste liquor, should be examined for the occurrence of antibiotic activity.

8. Conclusion

This survey, although far from complete, serves to show that stable mixed cultures of yeasts and lactic acid bacteria are widespread in a variety of seemingly unrelated environments. They are apparently able to outgrow, dominate and even destroy microbes which were originally present in much greater numbers than were the yeasts and lactic acid bacteria. In so doing they cause changes which can be used for the benefit of humanity and in consequence they have been so employed in diverse ways for many centuries.

Acknowledgement

My sincere thanks to Mrs J. Winter, Mrs Margaret Black and Mrs Anne Lawson, for producing an excellent typescript at very short notice.

References

Abiose, S.H. (1980). *Studies in Miso Fermentation*. Ph.D. Thesis, University of Strathclyde, Glasgow, Scotland.

Abiose, S.H., Allan, M.C. and Wood, B.J.B. (1981). Microbiology and biochemistry of *Miso* (soy paste) fermentation. Advances in Applied Microbiology (in press).

Akinrele, I.A., Cook, A.S. and Holgate, R.A. (1963). The manufacture of Gari from cassava in Nigeria. In "Food Science and Technology" (Ed. J.M. Leitch), Vol.IV, pp.633-44. London: Gordon and Breach, Science Publishers.

Aritaki, S. and Ishikawa, S. (1962). Application of *Lactobacillus acidophilus* fermented milk in the paediatric field. *Acta Paedriactica (Japan)* **66**, 811-15. Cited in *Dairy Science Abstracts* **26**(3): 745.

Bennion, E.B. (1954). Breadmaking, its principles and Practice, pp.119-22. Oxford: Oxford University Press.

Cook, A.S. and Holgate, R.A. (1963). The manufacture of gum from cassava in Nigeria. In "Food Science and Technology,4" (ed. J.M. Leith), pp.633-644. London: Gordon and Breach.

Dolan, T.C.S. (1979), Bacteria in whisky production. *The Brewer*, February 1979, 60-64.

Hamdan, I.Y. and Mikolajcik, E.M. (1974). Actidolin; an antibiotic produced by *Lactobacillus acidophilus*. *Journal of Antibiotics* **27**, 631-36. Cited in *Dairy Science Abstracts* **37**(167): 2743.

Hrubant, G.R. (1973). Characterization of the dominant aerobic microorganism in cattle feedlot waste. *Applied Microbiology* **26**, 512-16.

Hrubant, G.R. (1975). Changes in microbial population during fermentation of feedlot waste with corn. *Applied Microbiology* **30**, 113-19.

Hrubant, G.R., Daugherty, R.V. and Rhodes, R.A. (1972). Enterobacteria in feedlot waste and runoff. *Applied Microbiology* **24**, 378-83.

Hrubant, G.R. and Detroy, R.W. (1980). Composition and fermentation of feedlot wastes. In "Waste Treatment and utilization" (eds. M. Moo-Young and G.J. Farquhar), pp.411-23. Oxford and New York: Pergamon Press.

Hrubant, G.R., Rhodes, R.A. and Sloneker, J.H. (1978). Specific composition of representative feedlot wastes: A chemical and microbial profile. Science and Education administration, U.S. Dept. of Agriculture, Report No. SEA-NC-59. 94 pp.

Bryan-Jones, G. (1975). Lactic acid bacteria in distillery fermentations. In "Lactic acid bacteria in beverages and foods" (eds. J.G. Carr, C.V. Cutting and G.C. Whiting), pp.165-76. London and New York: Academic Press.

Kline, L. and Sugihara, T.F. (1971). Microorganisms of the San Francisco sourdough bread process. II. Isolation and characterization of undescribed bacterial species responsible for the souring activity. *Applied Microbiology* **21**, 459-65.

Kline, L. and Sugihara, T.F. (1973). Sour Dough French Bread. USA Patent No. 3,734, 743.

Kline, L., Sugihara, T.F. and McCready, L.B. (1970). Nature of the San Fransisco sourdough French bread process, I. Mechanism of the process. *Bakers' Digest* **44**, 48-50.

Kosikowski, F.V. (1977). Cheese and Fermented Milk foods (2nd edn.), pp.40-46. Ann Arbor, Michigan: Edwards Brothers.

Lloyd Hind, M. (1950). Brewing Science and Practice, Vol.II, pp.647-8. London: Chapman and Hall.

Novellie, L. (1980). Sorghum beer and related fermentations of Southern Africa. Paper presented to the VIth International Fermentation Symposium, London, Ontario, Canada.

Rhodes, R.A. and Hrubant, G.R. (1972). Microbial population of feedlot waste and associated sites. *Applied Microbiology* **24**, 269-77.

Rhodes, R.A. and Orton, W.L. (1975). Solid substrate fermentation of feedlot waste combined with feed grains. *Transactions of the American Society of Agricultural Engineers* **18**, 728-33.

Shahani, K.M., Vakil, J.R. and Kilara, A. (1977). Natural antibiotic activity of *Lactobacillus acidophilus* and *bulgaricus*, II. Isolation of Acidophilin from *L. acidophilus*. *Cultured Dairy Products Journal* **12**, 8-11. Cited in *Dairy Science Abstracts (1978)* **40**, 467.

Sloneker, J.H., Jones, R.W., Griffin, H.L., Eskins, K., Bucher, B.L. and Inglett, G.E. (1973). Processing animal wastes for feed and industrial products. In "Symposium; Processing Agricultural

and Municipal Wastes" (ed. G.E. Inglett), pp.13-28. Westport, Conn., USA: Avi Publishing Co.
Spicher, G. (1959). Die mikroflora des sauerteiges, 1. Untersuchungen über die art in sauerteigen anzutreffenden stäbchenförmigen milchsaurebakterien (genus Lactobacillus Beikerinck). *Zentralblatt für Bakteriologie, Parasitenkunde, Infektionskrankheiten und Hygiene, II* **113**, 80-106.
Spicher, G. (1974). Brot und andere backwaren. In "Ullmanns Encyklopädie der Technischen Chemie, 4. Neubearbeitete und Erweiterte Auflage, Band 8, Antimon bis Brot", pp.702-30. Weinheim: Verlag Chemie.
Spicher, G. and Schröder, R. (1978). Die mikroflora des sauerteiges, IV. Untersuchungen über die art in "Reinzuchtasuern" anzutreffenden stäbchenförmigen milchsäurebakterien (genus *Lactobacillus* Beijerinck) *Zeitschrift für Lebensmittel-Untersuchung und-Forschung* **167**, 342-54.
Spicher, G. and Schroder, R. (1979a). Die mikroflora des sauerteiges, V: Das vitaminbedurfnis der in "Reinzuchtsauern" und in sauerteigen anzutreffenden stäbchenförmigen milchsäurbakterien (genus *Lactobacillus* Beijerinck). *Zeitschrift für Lebensmittel-Untersuchung und-Forschung* **168**, 188-92.
Spicher, G. and Schroder, R. (1979b). Die mikroflora des sauerteiges, VI. Das aminosäurebedürfnis der in "Reinzuchtsauern" und in sauerteigen anzutreffenden stäbchenförmigen milchsäurbakterien (genus *Lactobacillus* Beijerinck). *Zeitschrift für Lebensmittel-Untersuchung und-Forschung* **168**, 397-401.
Spicher, G. and Schroder, R. (1980). Die mikroflora des saurteiges, VIII. Die faktoren des wachstums der in "Reinzuchtsauern" auftretenden hefen. *Zeitschrift für Lebensmittel-Untersuchung und-Forschung* **170**, 119-23.
Spicher, G., Schröder, R. and Schöllhammer, K. (1979). Die mikroflora des sauerteiges, VII. Untersuchungen über die art der in "Reinzuchtsauern" auftretenden hefen. *Zeitschrift für Lebensmittel-Untersuchung und-Forschung* **169**, 77-81.
Spicher, G., Schröder, R. and Stephan, H. (1980). Die mikroflora des saurteiges, X. Die backtechnische wirkung der in "Reinzuchtsauern" auftretenden milchsäurebakterien (genus *Lactobacillus* Beikerninck). Zeitschrift für Lebensmittel-Untersuchung und-Forschung (in press).
Spicher, G. and Stephan, H. (1964). Die mikroflora des sauerteiges, II. Untersuchungen über die backtechnische bedeutung der aus sauerteigen isolierten stäbchenförmigen milchsäurebakterien (genus *Lactobacillus* Beikerinck). *Zentralblatt für Bakteriologie, Parasitenkunde. Infektionskrankheiten und Hygiene II* **118**, 453-71.
Spicher, G. and Stephan, H. (1966). Die mikroflora des sauerteiges, III. Untersuchungen über die art der in "spontansauerteigen" anzutreffenden milchsäurebakterien und ihre backtechnische bedeutung. *Zentralblatt für Bakteriologie, Parasitenkunde, Infektionskrankheiten und Hygiene, II* **120**, 685-702.
Sugihara, T.F., Kline, L. and McCready, L.B. (1970). Nature of the San Fransisco sour dough French bread process, II. Microbiological aspects. *Bakers' Digest* **44**, 50-52.
Sugihara, T.F., Kline, L. and Miller, M.W. (1971). Microorganisms of the San Fransisco sour dough bread process, I. Yeasts

responsible for the levening action. *Applied Microbiology* **21**, 456-58.
Weiner, B.A. (1977a). Characteristics of aerobic, solid-substrate fermentation of swine waste-corn mixtures. *European Journal of Applied Microbiology* **4**, 51-57.
Weiner, B.A. (1977b). Fermentation of swine caste-corn mixtures for animal feed; pilot plant studies. *European Journal of Applied Microbiology* **4**, 59-65.
Weiner, B.A. and Rhodes, R.A. (1974). Growth of indigenous organisms in aerated filtrate of feedlot waste. *Applied Microbiology* **28**, 448-51.
Wood, B.J.B. (1977). Oriental food uses of *Aspergillus*. In "Genetics and Physiology of Aspergillus" (eds. J.E. Smith and J.A. Pateman), pp.481-98. London and New York: Academic Press.
Wood, B.J.B., Cardenas, Olga S., Yong, F.M. and McNulty, D.W. (1975). Lactobacilli in production of soy sauce, sourdough bread and Parisian barm. In "Lactic Acid Bacteria in Beverages and Food" (eds. J.G. Carr, C.V. Cutting and G.C. Whiting), pp.325-35. London and New York: Academic Press.
Wood, B.J.B. and Rainbow, C. (1961). The maltophosphorylase of beer lactobacilli. *Biochemical Journal* **78**, 204-9.
Wood, B.J.B. and Yong, F.M. (1975). Oriental food fermentations. In "The Filamentous Fungi" (eds. J.E. Smith and D.R. Berry), Vol. I. pp.265-80. London: Edward Arnold.
Yong, F.M. (1971). *Studies on Soy Sauce Fermentation*. M.Sc. Thesis, University of Strathclyde, Glasgow, Scotland.
Yong, F.M. and Wood, B.J.B. (1974). Microbiology and biochemistry of the soy sauce fermentation. *Advances in Applied Microbiology* **17**, 157-94.
Yong, F.M. and Wood, B.J.B. (1975). Sucrose from soy sauce moulds. *Transactions of the British Mycological Society* **64**, 1-3.
Yong, F.M. and Wood, B.J.B. (1976). Microbial succession in experimental soy sauce fermentations. *Journal of Food Technology* **11**, 1-12.
Yong, F.M. and Wood, B.J.B. (1977a). Biochemical changes in experimental soy sauce koji. *Journal of Food Technology* **12**, 163-75.
Yong, F.M. and Wood, B.J.B. (1977b). Biochemical changes in experimental soy sauce Moromi. *Journal of Food Technology* **12**, 263-73.

Chapter 8
THE USE OF ALGAL-BACTERIAL MIXED CULTURES
IN THE PHOTOSYNTHETIC PRODUCTION OF BIOMASS

Y.-K. LEE

*Microbiology Department, Queen Elizabeth College,
University of London, Campden Hill Road, London, UK*

1. Introduction

Algal cultivation has gained importance in recent years
as a means of harnessing and storing solar energy
[Benemann et al., 1977; Pirt et al., 1980] and as a
source of food, feedstuff and chemicals [Vincent, 1971;
Sanderson et al., 1978; Richmond and Preiss, 1980]. It
has also been adopted in industrial and domestic waste
treatment [Humenik and Hanna, 1971; Oswald, 1973] and in
environment control [Gale and Wixon, 1979]. Recently,
Horikoshi et al. [1979] reported on the ability of some
marine algae in extracting uranium from sea water.
    Phytoplankton can liberate a part of the carbon fixed
in photosynthesis into the external environment in the
form of dissolved organic materials, which serve a
variety of biological roles [Aaronson, 1973]. Among
these dissolved organic materials are polysaccharides
[Bishop et al., 1954], organic acids [Chang and Tolbert,
1970], polypeptides and amides [Fogg, 1952], amines
[Herrmann and Jüttner, 1977] and vitamins [Aaronson et
al., 1977]. Heterotrophic organisms may consume these
organic materials and flourish among the planktonic
population. Bacterial contamination of algal mass cultures was therefore not uncommon, especially in old culture plants which had been in use for long periods
[Gummert et al., 1953; Krauss and Thomas, 1954; Ward
et al., 1964; Blasco, 1965]. While investigating
bacterial contaminated algal culture, Blasco [1965] noted
a characteristic association between poor algal growth
and the presence of bacterial contaminants. Other
workers observed that contamination of algal culture by
pathogenic bacteria could wipe out the whole algal culture
population in a short time [Stewart and Brown, 1970;
Daft and Stewart, 1971; Bhumiratana and Payer, 1973].
Chemical control of bacterial outbreaks may be effective

[Galloway and Krauss, 1959; Blasco, 1963; Myers, 1963], but it is not desirable if the bacteria are metabolizing compounds that become autotoxins in algal culture [Pratt, 1944; Harris, 1971]. On the other hand, symbiotic algal-bacterial relationships have been reported in the literature [Kain and Fogg, 1958; Johnston, 1963] and the control of plant infections from pathogenic bacteria by antagonistic microorganisms is not a new subject in plant protection [Henis and Chet, 1975; Chattopadhay and Bose, 1980]. It is the intention of the present chapter to reveal the feasibility and practicality of using defined symbiotic algal-bacterial mixed cultures in the production of biomass. The term symbiotic is used here to designate beneficial associations, which include commensalistic and mutualistic relations between algal and bacterial species in a mixed culture. The use of crude, undefined mixed culture populations in biological waste treatment processes is beyond the scope of the present article.

2. Selection and Isolation of Mixed Cultures

The traditional batch culture enrichment technique is still useful in the preliminary screening of algal-bacterial mixed cultures from natural sources (water, soil, etc). Based on a simple selection parameter, such as maximum specific growth rate or biomass growth yield from nitrogen source or from light energy, several apparently superior mixed cultures may eventually be evolved for further selection. We considered that a selected algal-bacterial mixed culture would give optimum performance only if it had been selected under environmental conditions specifically designed for the desired application. A continuous chemostat or turbidostat culture fulfills the above purpose in providing a model system [Williams, 1971; Pirt, 1975; Harder et al., 1977]. The continuous enrichment technique has been widely used in the selection of desirable mixed microbial populations for industrial fermentation processes [Harder et al., 1977; Harrison, 1978].

The growth of a microbial culture limited by the energy source can be represented by [Pirt, 1975]:

$$\mu = q_m Y_G \frac{s}{K_m + s} - m Y_G \qquad (1)$$

where μ = specific growth rate, q_m = specific rate of limiting substrate utilization, Y_G = true growth yield, s = substrate concentration, K_m = saturation constant of limiting substrate, m = maintenance coefficient and [Pirt, 1965]:

$$1/Y = 1/Y_G + m/\mu \qquad (2)$$

where Y = actual growth yield

From Equation (1), the high selection pressure of continuous culture on growth rate and limiting substrate affinity [Pirt, 1975] will only select for symbiotic mixed cultures of high q_m and Y_G but low m. Presence of an unfavourable bacterium in the algal-bacterial mixed culture will lead to washout of the culture. As the actual growth yield is the primary process parameter for cost effective production of algal biomass, selection for high Y_G and low m is of paramount importance (Equation 2), and one may sacrifice other parameters, such as q_m. If we consider two algal cultures of the same maximum specific growth rate (μ_m), but with different values for Y_G, the culture with the higher Y_G (which we screen for) will have a lower q_m according to the relation $\mu = q_m Y_G$. A comparison of growth parameters of a selected algal-bacterial consortium MA003 [Lee and Pirt, 1979] and those from *Chlorella* 211/8k is presented in Table 1. The *Chlorella* 211/8k is a well established, high-temperature algal species widely used in the study of algal photosynthesis and mass cultivation [Sorokin and Myers, 1953; Sorokin, 1959]. Consortium MA003 has a higher Y_G value than *Chlorella* 211/8k but a much lower q_m, and consequently a lower maximum specific growth rate ($\mu = q_m Y_G$). A high biomass production rate (R) could nevertheless still be achieved by increasing the biomass concentration (x), since $R = \mu x$.

In order to study the interactions of the organisms present in an algal-bacterial mixed culture, each constituent has to be isolated in axenic culture, identified and studied. Thus one can obtain knowledge of the constituent organisms and reconstruct mixed populations by pairwise, or more complex, combinations. Meanwhile, in order to clarify, in the mixed cultures, the sources of N and C for bacterial growth, as well as studying the effect of extracellular products of one organism on the other, each constituent in mixed culture can be grown in the supernatant of the others, or in cultures separated by dialysis membrane. The difficulty in the isolation of axenic algal culture is well reflected by the great variety of isolation methods reported in the literature. We have successfully separated algae from algal-bacterial mixed cultures through streaking on nutrient plates. Algal nutrient plates may need to be enriched by growth factors, which otherwise are supplied by the coexisting bacterial species [Kain and Fogg, 1958; Johnston, 1963]. Unfortunately, algal plates usually require long incubation periods of up to three weeks before isolations can be made from them. Moreover plating is limited to the range of algae able to grow on solid media, such as *Chorophyta, Cyanophyta, Eugleminae* and pennate diatoms [Droop, 1969]. With some patience and practice, micromanipulation [Pringsheim, 1946; Droop, 1954] is probably the most successful isolation method which does not involve physical and chemical destruction of cells. In

TABLE 1

Growth Parameters of Photosynthetic Algal Cultures Grown at 37°C

	Y_G g dry biomass kJ^{-1}	m J g^{-1} dry biomass h^{-1}	q_m kJ g^{-1} dry biomass h^{-1}	μ_m h^{-1}
Consortium MA003	2.0	30	0.085	0.178
Chlorella vulgaris (Sorokin strain)	1.5	20	0.147	0.220

Y_G = true growth yield from light energy
m   = maintenance coefficient
q_m = maximum specific rate of light absorption
μ_m = maximum specific growth rate

this method, large dilution factors can be achieved by
transferring single cells through successive baths of
sterile medium with a micropipette. The method is limited
by the cell size, the lower limit being 5 µm [Droop,
1954]. Other mechanical manipulation techniques include
filtration [Heaney and Jaworski, 1977], equilibrium
centrifugation [Sitz and Schmidt, 1973] and atomizer
inoculation [Weinmann et al., 1964]. Techniques making
use of the gliding movements and phototaxis of motile
Cyanobacteria [Allen, 1973] have been reported. Vaara
et al. [1979] improved this isolation technique by
scoring the surface of algal plates with parallel lines.
The inoculated plates were incubated under unidirectional
light with the scores parallel with incident light.
After overnight incubation the cyanobacterial filaments
glided some distance from the inoculation point, along
the scores towards the incident light. This led to a
rapid separation of cyanobacterial filaments from their
adhering contaminants. The axenic filaments could be
picked out on an agar block and subcultured on fresh agar
plates. Twenty-two axenic cultures of cyanobacterial
strains belonging to the *Pseudoanabaena, Oscillatoria,
Lyngbya-Pnormidium-Plectonema* group and *Anabaena* were
established by this technique.

Many algae are more resistant than bacteria to antibiotics and chemicals, so they can be isolated in a
culture medium containing these agents. Combinations of
penicillin, ceporin, anreomycin, tetracycline, chloramphenicol, aureomycin, streptomycin and neomycin have been
successfully used in the isolation of many diatoms and
green algae [Jones et al., 1973]. Simplicity and reliability have made the antibiotic technique widely used in
routine maintenance of axenic culture stocks of algae.
The drawback is that, after some time, antibiotic resistant bacteria appear in the cultures. Chemical agents
for removing contaminant bacteria from algal cultures
include chlorine water [Fogg, 1942] and phenol [McDaniel
et al., 1962]. Phenol when added to dark treated algal
cultures selectively reduced the number of actively
growing bacteria, but left the resting algal cells viable
[Carmichael et al., 1974]. The use of ultraviolet treatment in isolating Myxophyceae was first introduced by
Allison and Morris [1930]. Algal cultures were suspended
in a quartz test tube and irradiated for 15 to 30 min
with 275 nm ultraviolet light from a mercury vapour lamp.
Aliquots were removed and an appropriate time interval
produced viable algae free from bacteria. On the other
hand Krauss [1966] isolated ten species of blue green
algae free from bacterial contaminants by irradiating
the culture with γ-rays from a ^{60}Co source. Treatment of
cyanobacterial cultures with heat [Wieringa, 1968] was
claimed to be successful in removing bacterial
contaminants.

3. Algal-Bacterial Interactions in Mixed Cultures

In an algal-bacterial culture growing on carbon dioxide, light energy and minimal chemical medium, the first obvious benefits which heterotrophic bacteria would receive from photosynthetic algal cells are the supply of fixed carbon and oxygen. Vela and Guerra [1966] showed that bacterial proliferation in mixed cultures of *Chlorella* was a function of algal growth. By feeding axenic *Chlorella* culture [^{14}C]-labelled bicarbonate under laboratory culture with illumination, McFeters *et al.* [1978] obtained radioactive algal products in the culture supernatant. These became incorporated into coliform bacteria as they grew, and were later released when they died. Blasco [1965] however, observed that products released by some *Chlorella* species did not provide sufficient C-source for growth of coexisting bacteria. The heterotrophic bacteria might therefore be feeding on the peripheral sheath of the *Chlorella* cells or on cell wall material released during cell division. Bacteria obtaining growth substrates from photosynthetic algae through a parasitic mode of action were also reported [Daft and Stewart, 1971]. These substances, capable of supporting bacterial oxidation and growth, were varied in kind and were utilized selectively by different bacteria [Vela and Guerra, 1966; Bell *et al.*, 1974]. Consequently, not all of the soil and airborne bacteria can proliferate in mixed cultures with algae or in filtrates of the axenic algal cultures [Vela and Guerra, 1966]. The bacteria commonly isolated from mass cultures of green algae growing between 25°C and 37°C were *Flavobacterium, Pseudomonas, Mima, Aerobacter, Bacillus, Bacterium, Staphylococcus, Micrococcus* and *Serratia* [Krauss and Thomas, 1954; Blasco, 1963; Mayer *et al.*, 1964; Ward *et al.*, 1964; Lee and Pirt, 1979]. The heterotrophic bacteria commonly associated with cyanobacteria in mixed cultures were *Pseudomonas* and *Caulobacter* [Bunt, 1961; Gromov, 1964; Bershova *et al.*, 1968].

Myers *et al.* [1951] stated that the presence of bacteria had little effect on the growth rate of *Chlorella pyrenoidosa*. Several years later in 1963, Myers also reported that bacterial contaminants usually had no effect on algae in rapidly growing steady state cultures. Thus the heterotrophic bacteria appeared merely as scavengers in algal cultures. The same was observed by Mayer *et al.* [1964] on mass cultures. On the other hand, Nakamura [1963] found that selected bacterial species enhanced the growth of algae in nutritionally deficient cultures. He postulated that mixed cultures of *Chlorella* and bacteria form an active symbiotic relationship both in nature and in laboratory cultures. Vance [1966] showed that a strain of *Microcystis aeruginosa* failed to grow in the absence of associated bacteria. Other evidence of close interrelations between planktonic algae and their associated bacteria came from studies of marine algae by Kain and Fogg [1958] and Johnston [1963]. Kain and Fogg [1958] reported that *Asterionella japonica* grew satisfactorily in culture as long as bacteria were present, but ceased to grow when the bacteria were removed, even though a variety of possible organic growth factors were provided. They claimed that this effect was due to production of vitamin B12 by the bacteria. Johnston [1963] also found that bacteria-free *Skeletonema costatum* grew poorly in various samples of sea water enriched with essential minerals, but grew distinctly better in the same media if selected bacteria were present. Among the nitrogen-fixing cyanobacteria, Bunt [1961] and Bjälfve [1962] found that the presence of some bacteria increased nitrogen-fixation by *Nostoc* species, even though the bacteria were themselves incapable of fixing nitrogen. It was suggested that at least part of the effect may have been associated with a reduction in oxygen concentration, which in some cyanobacteria increased the growth rate [Gusev, 1962], carbon dioxide fixation and nitrogen fixation [Stewart and Pearson, 1970].

Contrary to this, during the course of studies on mass culture of *Chlorella pyrenoidosa*, Sorokin and Myers [1953] and Ward et al. [1964] noted a characteristic correlation between poor algal growth and the presence of bacterial contaminants. Blasco [1965] also observed subnormal rates of growth and photosynthetic gas exchange when bacteria were found imbedded in the surface of *Chlorella* cells. He reasoned that since coexisting bacteria did not impose direct effects on the net algal photosynthetic gas exchange, the possible role of these contaminants was pathogenic. Blasco identified the bacteria isolated from mass algal cultures as *Pseudomonas* sp. and *Xanthomonas* sp.. Bershova et al. [1968] isolated strains of bacteria growing among cyanobacteria and reported that one quarter of all strains studied were antagonistic to *Microcystis pulverea*. However, the "satellite" bacteria isolated from the mucilages of the algae showed few antagonistic species. It is however not clear whether it was those algal cells which carried nonantagonistic satellite bacteria which proliferated, or those harmless satellite bacteria which protected the algae from the antagonistic effect of other bacteria.

The reported ratio of algae to bacteria in mixed cultures varies widely, and possibly depends on strain and state of culture organisms. Ward et al. [1964] stated that the viable bacterial population contributed less than 1% of the *Chlorella* culture mass. However Jones [1972] and Taub [1969] estimated that the ratio of algal and bacterial biomass was 60 to 1. An even higher ratio of 4 to 1 for algal and bacterial biomass was reported by Mayer et al. [1964].

The few reports on relationships between algae and

bacteria in a mixed culture mentioned above cover the
whole spectrum of microbial interactions, but most
consider binary systems. In practice, the interactions
between the algal and bacterial species in a stable mixed
culture are multi-lateral and complex. Lee and Pirt
[1979, 1981] have isolated a high temperature (37°C)
consortium of alga and bacteria using the continuous
enrichment method. The consortium MA003 consisted of
Chlorella-like green alga and three heterotrophic bacteria,
Alcaligenes sp., *Flavobacterium* sp. and *Serratia* sp.. In
terms of biomass, the algal species dominated the four-
membered consortium MA003 (> 80%) in both N and light
energy limited chemostat cultures. On the other hand,
the bacterial population ratio remained as 2000:200:1 for
Alcaligenes, *Flavobacterium* and *Serratia* in chemostat
operating at dilution rate between $0.02\ h^{-1}$ and $0.18\ h^{-1}$.
Their interactions in the consortium MA003 are depicted
in Fig. 1. Given that the three bacteria were not photosynthetic and did not appear to use urea nitrogen in the
culture medium, the carbon and nitrogen substrates for
the bacteria were probably obtained from algal cells in

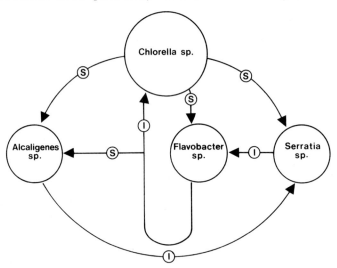

Fig. 1. Schematic representations of the interactions between the
four members of the algal-bacterial consortium MA003. S: substrates,
I: inhibition, the arrows indicate direction of action.

the mixed culture. It was further noted in batch culture
experiments that the bacteria attained maximum population
densities in urea depleted axenic algal culture supernatant. It was therefore assumed that the carbon and
nitrogen substrates for bacterial metabolism were organic
carbon and nitrogen compounds excreted by the alga.
Algal maximum specific growth rate was however greatly
decreased ($0.17\ h^{-1}$ to $0.07\ h^{-1}$) when growing alone with t

Flavobacterium sp. in chemostat cultures, and by the culture supernatant of the bacterium. The inhibiting effect suggested the production of a soluble algal inhibitory substance by the *Flavobacterium*. *Alcaligenes* sp. on the other hand appeared to consume this inhibitor and increased in steady state biomass in three-membered mixed cultures with the alga and the *Flavobacterium*. The steady state biomass of *Alcaligenes* was much lower (1/3) when growing alone with the alga in the chemostat cultures. The *Serratia* sp. decreased the growth of *Flavobacterium* sp. and consequently the production of algal growth inhibitor. This ability of *Serratia* contributed little in the consortium MA003, where the growth of *Serratia* sp. was greatly suppressed probably by the bacterium *Alcaligenes*. The system shows that the adverse effect of a bacterium (*Flavobacterium* sp.) on an algal culture was removed by the coexisting bacteria (*Alcaligenes* sp. + *Serratia* sp.) present in the system.

With urea-limited growth, there was a small increase in the growth yield from urea nitrogen (12%) in consortium MA003 over the single membered algal culture (Table 2).

TABLE 2

The Effect of Respective Coexisting Bacteria on the Maximum Specific Growth Rate (μ_m), Growth Yield from Urea Nitrogen (Y_N) and Growth Yield from Light (Y_G) of the Photosynthetic Chlorella sp. in the Consortium MA003. The Axenic Chlorella Culture and Various Mixed Cultures Were Grown in Minimal Chemical Defined Medium in Chemostat at 37°C.

Cultures	μ_m (h^{-1})	Y_N (g dry biomass $g^{-1}N$)	Y_G (g dry biomass KJ^{-1})
Consortium MA003	0.178	10.717	0.0206
Chlorella sp	0.176	9.410	0.0206
Chlorella + *Alcaligenes*	0.172	9.410	-
Chlorella + *Flavobacter*	0.058	10.383	-
Chlorella + *Serratia*	0.170	9.734	-
Chlorella + *Alcaligenes* + *Flavobacter*	0.180	10.383	-
Chlorella + *Alcaligenes* + *Serratia*	0.155	10.058	-
Chlorella + *Flavobacter* + *Serratia*	0.170	10.545	-

The coexisting bacteria *in toto*, in consortium MA003 did not significantly affect the algal maximum specific growth rate (μ_m) and algal growth yield from light energy (Y_G), up to a culture density of about 36 g l^{-1} (Table 2).

4. Energy and Mass Balance of Mixed Cultures

One approach to the study of the quantity of fixed carbon excreted by photosynthetic algae is to use radioactive markers [Fogg et al., 1965; Anderson and Zeutschel, 1970; Thomas, 1971]. By using radioactive carbon dioxide, Fogg et al. [1965] found that fresh water and marine phytoplankton excreted 7 to 50% of the radioactivity taken up by the algae back into their environment, thus the algae contributed a substantial amount of the organic materials in the aquatic environment. Anderson and Zeutschel [1970] also reported that phytoplankton labelled with radioactive markers in marine inlets lost 1 to 49% of the [^{14}C]-carbon dioxide fixed to the water as organic molecules. Again, Thomas [1971] observed that 7 to 44% of the carbon photoassimilated by marine algae was released as dissolved organic matter.

Our alternative to this is to use data from an energy balance approach. Thus the balance for energy source utilization in a light energy limited photosynthetic microbial culture is given by Pirt [1965],

$$q = \mu/Y_G + m \qquad (3)$$

In the case of the algal-bacterial mixed culture growing on light energy and carbon dioxide, the algal species was the sole utilizer of the light energy and primary carbon. The heterotrophic bacteria present in the mixed culture depended on carbon fixed by algae for growth. The heterotrophs in the algal-bacterial mixed culture could therefore be considered as part of the integral photosynthetic culture. Algal carbon and energy products consumed by bacteria in a mixed culture could be produced at a constant rate (J(g dry biomass)$^{-1}$h^{-1}) irrespective of algal specific growth rate, or at a constant yield (J(g dry biomass)$^{-1}$).

If cellular matter is excreted from growing photosynthetic algal cells at a constant rate independent of specific growth rate, the rate of light utilization for this product synthesis (q_p) can be incorporated into the maintenance term of the energy balance of algal cells:

$$q_A = \mu_a/Y_{GA} + q_p + m_a \qquad (4)$$

where q_A = specific rate of energy absorption by algal cells, μ_a = specific growth rate of algal cells, Y_{GA} = true growth yield from light energy of algal cells alone, q_p = specific rate of energy substrate utilization for extracellular product formation,

m_a = maintenance coefficient of algal cells. In a photosynthetic algal-bacterial mixed culture, if the bacteria consumed products are excreted by the algal cells for biomass production, then,

$$q_T = \mu_T/Y_{GB} + m_T$$

where q_T = specific rate of energy absorption by algal and bacterial cells, μ_T = specific growth rate of algal and bacterial mixed culture, Y_{GB} = true growth yield from light energy of the algal-bacterial mixed culture, m_T = maintenance coefficient of algal-bacterial mixed culture

if $m_T \simeq m_a$

then $q_T = \mu_T/Y_{GB} + m_a$ \hfill (5)

At steady state of chemostat culture, $\mu_a = \mu_T = D$, where D is the dilution rate.

when D = 0, and Eq.4 - Eq.6,

$$q_A - q_T = q_p \hspace{2cm} (6)$$

For the consortium MA003, it is shown in Fig. 2 that, $Y_{GA} = Y_{GB} = 2.0 \times 10^{-5}$ g J^{-1}. And $q_p = 418-30 = 388$ J(g dry

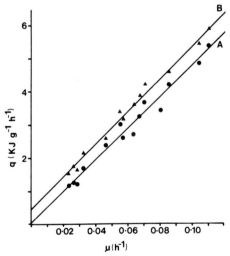

Fig. 2. Specific light absorption rate (q) as a function of specific growth rate (μ) in steady state of light-limited chemostat culture of consortium MA003. Line A for total biomass (algal + bacterial), line B for algal biomass only (total biomass-bacterial biomass). [Pirt et al., 1980].

biomass)$^{-1}$ h^{-1}. Also Fig. 2 shows that m_a was minute (30 J(g dry biomass)$^{-1}$ h^{-1}), but $(m_a + q_p)$ was significant (418 J(g dry biomass)$^{-1}$ h^{-1}), and represented 4% of the energy uptake by algal cells growing at maximum specific growth rate. It seems that, in a light limited culture, the heterotrophic bacteria in the consortium scavenged most of the carbon and energy products which were excreted at a constant rate (388 J(g dry biomass)$^{-1}$ h^{-1}) by the algal cells, and thus improved the total biomass production. Since there was a constant specific rate of production of extracellular algal products, the yield of these products varied with growth rate. This might explain the large variation in percentages of excreted fixed carbon compounds observed by those workers who used radioactive labelling [Fogg et al., 1965; Anderson and Zeutschel, 1970; Thomas, 1971].

Alternatively, the extracellular products of photosynthetic algae can be produced at a constant yield Y_p,

since $\quad q_p = Y_p \mu_a$

then Eq. 4 becomes $\quad q_A = \mu_a / Y_{GA} + Y_p \mu_a + m_a$

$$q_A = \mu_a(1 + Y_p Y_{GA})/Y_{GA} + m_a \tag{7}$$

Plots of specific rate of energy absorption (q) against specific growth rate (μ) are schematically presented in Fig. 3.

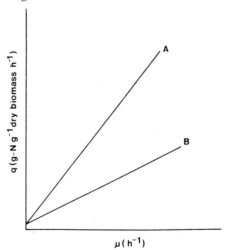

Fig. 3. Schematic representation of Equation 5 and Equation 7. Line A is the plot of q_A against μ_a in Eq. 7, line B is the plot of q_T against μ_T in Eq. 5. Y_p = constant.

The possibility of differentiating urea consumption by algal cells in consortium MA003 into growth linked and non-growth linked components in a nitrogen limited culture has also been investigated. The mathematical treatment of Equation 3, used to distinguish between the components of energy consumption, is used. Here, m is replaced by k, where k = non-growth linked metabolism of alga. Thus,

$$q = \mu/Y_N + k \qquad (8)$$

where Y_N = true yield from urea. Since k of consortium MA003 is constant, the plot of specific urea uptake rate, against μ is linear (Fig. 4), with slope $1/Y_N$ and intercept k on the ordinate. $Y_N = 5.3$ g dry biomass (g urea)$^{-1}$ is the same irrespective of whether the total real biomass (algae + bacteria) or just the algal biomass was used in the calculation of q. However k in each plot in

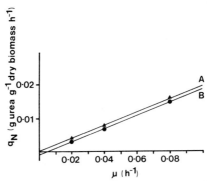

Fig. 4. Specific urea uptake (q) as a function of specific growth rate (μ) in steady states of a urea-limited chemostat culture of consortium MA003. Line A for total real biomass (algal + bacterial - starch), line B for real algal biomass only. [Lee and Pirt, 1981].

Fig. 4 differed. The difference between these k values (0.001 g urea (g dry biomass)$^{-1}$ h^{-1}) is taken to be the non-growth linked portion of the urea consumed by the algae and subsequently utilized by bacteria in the mixed culture for biomass growth. The specific rate of excretion of the algal N-products was therefore, $(0.001 \times 28)/(14 \times 60.06) = 33.3$ μmol N(g dry biomass)$^{-1}$ h^{-1}, and independent of algal growth rate.

5. Stability of Mixed Cultures

Our faith in algal-bacterial mixed cultures for biomass production lies in the microbiological stability of mixed cultures. By growing the established algal-bacterial consortium MA003 in a culture medium prepared from non-sterilized tap water or in medium exposed to aerial contamination, it was observed that the stability of the mixed culture was unaffacted, and no foreign organism appeared in the course of two months chemostat culture, with dilution rate at about 0.06 h^{-1}. Here the well established ecosystem of the alga and bacteria in consortium MA003 appeared to exclude foreign bacteria. We have to point out also that as Co^{2+} ion was excluded from the culture medium, prokaryote contamination was therefore restricted, since cobalt is required by prokaryotes in the synthesis of cobalamin.

Algal cultivation plants are characterized by the large surface area to culture volume ratio, due to the form of energy (light) arriving at the surface of a culture. As a result, culture environments such as temperature and pH of an algal culture plant cannot be stringently controlled, as is necessary in other microbial fermentation process. Laboratory experiments under controlled conditions were therefore undertaken to study the effect of these aspects on growth. Urea limited consortium MA003 was grown at a constant dilution rate of about half of its maximum specific growth rate (D = 0.066 h^{-1}) in chemostat. It was observed (Fig. 5) that the culture composition varied

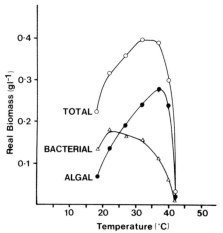

Fig. 5. Effect of temperature (°C) on steady state real biomass of alga (algal-starch),●, bacteria, △, and total real biomass (algal + bacterial-starch), O, of the consortium MA003 grown in a chemostat culture. Growth limiting substrate, urea = 0.067 g l^{-1}, dilution rate = 0.066 h^{-1}.

with the change in culture temperature. At higher temperatures (30-40°C) the culture was dominated by the alga with a maximal steady state biomass concentration at 37°C. Bacteria grew best at a wider but lower temperature range from 18-30°C, with an optimum at 22°C. Consequently, the total steady state biomass remained rather constant between 32°C and 37°C. The steady state total biomass dropped off sharply at higher temperature, but declined gradually at lower temperatures. On the other hand, the data for pH effects on consortium MA003 (Fig. 6) shows a curve in which maximum steady state algal biomass was achieved at pH 6.5, the acidity at which

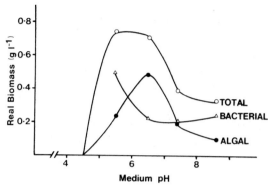

Fig. 6. Effect of medium pH value on steady state real algal biomass (algal-starch),●, bacterial biomass, △, and total real biomass (algal+bacterial-starch), ○, in the consortium MA003 grown in a chemostat. Growth limiting substrate, urea = 0.067 g l^{-1}, dilution rate = 0.066 h^{-1}.

MA003 was originally selected. The response of bacteria in consortium MA003 to pH changes was somewhat surprising, the bacterial population proliferating readily at pH value unfavourable for algal growth. They grew best in a slightly acidic environment. The resulting steady state total biomass had its maximum concentration at pH 5.5, and decreased with increasing pH values of culture medium. Apparently, when growing at extreme culture temperatures and pH values (Fig. 5 and Fig. 6), the alga in the consortium MA003 excreted increased amounts of N and C products. The various excreted algal products were subsequently consumed by the three bacterial species which proliferated in the mixed culture. The temperature and pH value for maximum population was different for each bacterium (*Alcaligenes* at 18°C, pH 7.4; *Flavobacterium* at 27°C, pH 5.5; *Serratia* at 33°C, pH 8.6). Nevertheless, the population ratio of each of the bacteria in the consortium was always of the same order of magnitude *Alcaligenes* > *Flavobacterium* > *Serratia*. The effect of diurnally intermittent illumination (24 h cycle) on the stability of consortium MA003 was also studied. In the 12 h dark period in batch, which was proceeded by a 12 h light energy limited chemostat growth, a drop in algal biomass was observed. Part of the reduction in algal biomass was due to the endogenous metabolism of resting cells [Pirt, 1975]. Release of cellular materials from algal cells, which in turn induced an average 25% increase in bacterial biomass over the dark period, also probably contributed to the reduction of algal biomass in the dark. As a result, the total (alga + bacteria) biomass production remained fairly constant. The cyclic changes in algal and bacterial biomass concentration repeated in each light-dark cycle.

6. Prospects for Mixed Cultures

The photosynthetic production of biomass for food and feedstuff is appealing, where there is a deficiency of both conventional energy and fixed carbon sources. The primitive unicellular characteristic of microscopic algae gives them a number of advantages over higher conventional crop production [Vincent, 1971; Bhattacharjee, 1970]. In terms of dry biomass yield, algal will give at least 15 times more than the best productivity from conventional agriculture [Pirt et al., 1980]. However, the most commonly used algal in mass culture studies such as *Chlorella* and *Scenedesmus* have low methionine contents (Table 3). Other essential amino acids for mammalian nutrient, such as valine, tryptophan and isoleucine are generally lower than would be desirable in a balanced protein diet according to the FAO standard (Table 3). Selection for suitable algal strains and mutants may solve the problem. On the other hand, the bacterial biomass protein could compensate for the amino acid deficiency of the algae [Wagner et al., 1969; Chalfan and Mateles, 1972; Gow et al., 1975]. Algae and bacteria have vastly different physiological and biochemical properties. Many of the complementary properties are certainly worth exploiting. Another possible beneficial association of algae and bacteria in mixed culture for biomass production comes from the reports that the extracellular materials of *Anabaena cylindrica* neutralize the antibiotic effect of polymyxin B [Whitton, 1965, 1967]. Polymyxin B was produced commercially from strains of *Bacillus polymyxa*, to which algae are otherwise relatively sensitive. Incorporation of a suitable strain of *Bacillus polymyxa* into an *Anabaena* system may therefore eliminate the possibility of contamination and take over by other algal species.

7. General Conclusion

In order to produce a well defined algal-bacterial consortium of favourable interactions, each organism in the selected mixed culture must be isolated, identified and studied. A defined consortium was finally evolved through reconstitution of the isolated organism and re-examination of the performance of the mixed culture. Nevertheless, the advantages of an algal-bacterial consortium over a unialgal culture for biomass production can be easily recognized:
(1) by occupying the ecological niches created by the

TABLE 3

Content of Essential Amino Acid of Algae and Bacteria

Organisms	Essential Amino Acid (%)								Reference
	Methionine	Valine	Tryptophan	Phenyl-alanine	Lysine	Leucine	Threonine	Isoleucine	
FAO Standard	2.2	4.2	1.4	2.8	4.2	4.8	2.8	4.2	
Chlorella sp.	0.36	2.47	0.80	1.81	2.06	2.79	2.12	1.75	Fisher and Burlew [1953]
Chlorella sp.	0.57	2.67	0.41	2.14	2.43	1.99	1.91	1.69	Krauss [1962]
Scenedesmus acutus	0.37	3.02	–	2.57	3.52	4.38	2.73	2.02	Bhumiratana [1977]
Nocardia sp.	3.2	4.7	–	2.9	4.0	6.7	4.7	3.7	Wagner et al., [1969]
Pseudomonas C	2.6	6.7	–	4.0	8.3	8.1	5.9	5.7	Chalfan and Mateles [1972]
Pseudomonas methylotropha	3.1	6.7	–	4.4	7.5	8.6	5.9	5.5	Gow et al., [1975]

photosynthetic algae, the coexisting bacteria prevent intrusion of other undesirable heterotrophic organisms. When intrusion to the mixed culture does occur, the existing bacteria in vigorous growth may exert the ability to neutralize or remove the deleterious factor to biomass growth;
(2) it provides a well defined culture population;
(3) by scavenging the excretion of algal cells as energy and nitrogenous sources for biomass growth, bacterial symbiosis actually increase the biomass production, especially when growth conditions are suboptimal. Meanwhile, it is also important to note that the N yield and energy yield from light of algal cells of a mixed culture are not necessarily affected by the presence of bacteria;
(4) bacteria may supplement the nutritional value of algal biomass in food and feedstuff production.

Acknowledgements

The author is deeply grateful to Professor John Pirt and John Hodgson for their advice and criticism during the preparation of this article.

References

Aaronson, S. (1973). *Ochromonas*: A model system for the study of phytoflagellate and algal secretion into the environment. In "Modern Methods in the Study of Microbial Ecology" (ed. T. Rosswal), pp.367-370. Swedish Natural Science Research Council.

Aaronson, S. Dhawale, S.W., Patni, N.T., DeAngelis, B., Frank, O. and Baker, H. (1977). The cell content and secretion of water-soluble vitamins by several fresh water algae. *Archives of Microbiology* **112**, 57-9.

Allen, M.M. (1973). Methods for Cyanophycaea. In "Handbook of Phycological Methods (ed. J.R. Stein), pp.127-132. Cambridge University Press.

Allison, F.E. and Morris, H.J. (1930). Nitrogen fixation by blue-green algae. *Science*, New York **71**, 221-3.

Anderson, G.C. and Zeutschel, R.P. (1970). Release of dissolved organic matter by marine phytoplankton in coastal and offshore areas of the northeast pacific ocean. *Limnology and Oceanography* **15**, 402-7.

Bell, W.H., Long, H.M. and Mitchell, R. (1974). Selective stimulation of marine bacteria by algal extracellular products. *Limnology and Oceanography* **19**, 833-9.

Benemann, J.R., Weissman, J.C., Koopman, B.L. and Oswald, W.J. (1977). Energy production by microbial photosynthesis. *Nature*, London **268**, 19-23.

Bershova, O.I., Kopteva, Zh.P. and Tantsyurneko, E.V. (1968). The Interrelations between the blue-green algae — the causative agents of water bloom — and bacteria. In "Tsvetenie Vodȳ" (ed. A.V. Topachevsky), pp.159-171. Kiev: Naukova Dumka.

Bhattacharjee, J.K. (1970). Microorganisms as potential sources of

food. *Advances in Applied Microbiology* **13**, 139-61.
Bhumiratana, A. (1977). Algae project. Bangkok: Institute of Food Research and Product Development, Kasetsart University.
Bhumiratana, A. and Payer, H.D. (1973). Algae project. Bangkok: Institute of Food Research and Product Development, Kasetsart University.
Bishop, C.T., Adams, G.A. and Hughes, E.D. (1954). A polysaccharide from the blue-green alga, *Anabaena cylindrica*. *Canadian Journal of Chemistry* **32**, 999-1004.
Bjälfve, G. (1962). Nitrogen fixation in cultures of algae and other microorganisms. *Physiologia Plantarum* **15**, 122-9.
Blasco, R.J. (1965). Nature and role of bacterial contaminants in mass culture of thermophilic *Chlorella pyrenoidosa*. *Applied Microbiology* **13**, 473-7.
Bunt, J.S. (1961). Nitrogen-fixing blue-green algae in Australian rice soils. *Nature*, London **192**, 479-80.
Carmichael, W.W. and Gorham, P.R. (1974). An improved method for obtaining axenic clones of planktonic blue-green algae. *Journal of Phycology* **10**, 238-40.
Chalfan, Y. and Mateles, R.I. (1972). New pseudomonad utilizing methanol for growth. *Applied Microbiology* **23**, 135-40.
Chang, W-H. and Tolbert, N.E. (1970). Excretion of glycolate, mesotartrate and isocitrate lactone by synchronized culture of *Ankistrodesmus braunii*. *Plant Physiology* **46**, 377-85.
Chattopadhay, J.P. and Bose, S.K. (1980). Control of plant infections by antibiotics and antagonistic organisms. *Process Biochemistry* **15**, 5, 27-8.
Daft, M.J. and Stewart, W.D.P. (1971). Bacterial pathogens of fresh water blue-green algae. *New Phytologist* **70**, 819-29.
Droop, M.R. (1954). A note on the isolation of small marine algae and flagellates for pure cultures. *Journal of the Marine Biology Association of the United Kingdom* **33**, 511-4.
Droop, M.R. (]969). Algae. In "Methods in Microbiology" 3B (eds. J.R. Norris and D.W. Ribbons), pp.269-313. London: Academic Press.
Fisher, A.W. Jr. and Burlew, J.S. (1953). Nutritional value of microscopic algae. In "Carnegie Institute of Washington publication 600: *Algal Culture*" (ed. J.S. Burlew), pp.303-310. Washington DC: Carnegie Institute of Washington.
Fogg, G.E. (1942). Studies on nitrogen fixation by blue-green algae, 1. Nitrogen fixation by *Anabaena cylindrica* Lemm. *Journal of Experimental Botany* **19**, 78-87.
Fogg, G.E. (1952). The production of extracellular nitrogenous substances by a blue-green alga. *Proceedings of the Royal Society of London* B **139**, 372-97.
Fogg, G.E., Nalewajko, C. and Watt, W.D. (1965). Extracellular products of phytoplankton photosynthesis. *Proceedings of the Royal Society of London* B **162**, 517-34.
Gale, N.L. and Wixon, B.G. (1979). Removal of heavy metals from industrial effluents by algae. *Developments in Industrial Microbiology* **20**, 259-74.
Galloway, R.A. and Krauss, R.W. (1959). The differential action of chemical agents, especially polymyxin B on certain algae, bacteria and fungi. *American Journal of Botany* **46**, 40-49.

Gow, J.S., Littlehailes, J.D., Smith, S.R.L. and Walter, R.B. (1975). SCP production from methanol: bacteria. In "Single Cell Protein II" (eds. S.R. Tannenbaum and D.C. Wang), pp.370-384. Massachucetts: MIT Press.

Gromov, B.V. (1964). Bacteria of the genus *Caulobacter* accompanying algae. *Mikrobiologiya*, Translation 33, 263-68.

Gummert, F., Meffert, M-E. and Stratmann, H. (1953). Non-sterile large scale culture of *Chlorella* in greenhouse and open air. In "Carnegie Institute of Washington publication 600: *Algal culture*" (ed. J.S. Burlew), pp.166-176. Washington DC: Carnegie Institute of Washington.

Gusev, M.V. (1962). Vliyanie rastworinnogo kisloroda na razvitie sinezelenyh vodoroseley. *Doklady Akademii Nauk SSSR* 147, 947-50.

Harder, W., Kuenen, J.G. and Matin, A. (1977). Microbial selection in continuous culture. *Journal of Applied Bacteriology* 43, 1-24.

Harris, D.O. (1971). Growth inhibitors produced by the algae (*Volvocaceae*). *Archiv für Mikrobiology* 76, 47-50.

Harrison, D.E.F. (1978). Mixed cultures in industrial fermentation processes. *Advances in Applied Microbiology* 24, 129-62.

Heaney, S.I. and Jaworski, G.H.M. (1977). A simple separation technique for purifying micro-algae. *British Phycological Journal* 12, 171-74.

Henis, Y. and Chet, I. (1975). Microbiological control of plant pathogens. *Advances in Applied Microbiology* 19, 85-111.

Herrmann, V. and Jüttner, F. (1977). Excretion products of algae. Identification of biogenic amines by gas-liquid chromatography and mass spectrometry of their trifluoroacetamides. *Analytical Biochemistry* 78, 365-73.

Horikoshi, T., Nakajima, A. and Sakaguchi, T. (1979). Uptake of uranium from sea water by *Synechococcus elongatus*. *Journal of Fermentation Technology* 53, 197-4.

Humenik, F.J. and Hanna, G.P. Jr. (1971). Algal-bacterial symbiosis for removal and conservation of waste water nutrients. *Journal of Water Pollution Control Federation* 43, 580-94.

Johnston, R. (1963). Sea water the natural medium of phytoplankton I. General features. *Journal of the Marine Biological Association of the United Kingdom* 43, 427-56.

Jones, J.G. (1972). Studies on fresh water bacteria: association with algae and alkaline phosphatase activity. *Journal of Ecology* 60, 59-75.

Jones, A.K., Rhodes, M.E. and Evans, S.C. (1973). The use of antibiotics to obtain axenic cultures of algae. *British Phycological Journal* 8, 185-96.

Kain, J.M. and Fogg, G.E. (1958). Studies on the growth of marine phytoplankton, I. *Asterionella japonica* Gran. *Journal of the Marine Biological Association of the United Kingdom* 37, 397-413.

Krauss, R.W. (1962). Mass culture of algae for food and other organic compounds. *American Journal of Botany* 49, 425-35.

Krauss, M.P. (1966). Preparation of pure blue-green algae. *Nature, London* 211, 301.

Krauss, R.W. and Thomas, W.H. (1954). The growth and inorganic nutrition of *Scenedesmus obliquus* in mass culture. *Plant Physiology* 29, 205-14.

Lee, Y-K. and Pirt, S.J. (1979). Interactions in an algal-bacterial mixed culture used in the photosynthetic production of biomass and starch. *Society of General Microbiology Quarterly* **6**, 92.

Lee, Y-K. and Pirt, S.J. (1981). Interactions between an alga and three bacterial species in a consortium selected for photosynthetic biomass and starch production. *Journal of Chemical Technology and Biotechnology* **31**, 295-305.

Mayer, A.M., Zuri, U., Shain, Y. and Ginzburg, H. (1964). Problems of design and ecological considerations in mass culture of algae. *Biotechnology and Bioengineering* **6**, 173-90.

McDaniel, H.R., Middlebrook, J.B. and Bowman, R.O. (1962). Isolation of pure cultures of algae from contaminated cultures. *Applied Microbiology* **10**, 223.

McFeters, G.A., Stuart, S.A. and Olson, S.B. (1978). Growth of heterotrophic bacteria and algal extracellular products in oligotrophic water. *Applied and Environmental Microbiology* **35**, 383-91.

Myers, J. (-963). Algal cultures. In "Kirk-Othmer Encyclopedia of Chemical Technology 1" (eds. H.F. Mark, J.J. McKetta Jr. and D.F. Othmer), pp.649-68. New York: John Wiley and Son, Inc.

Myers, J., Phillips, J.N. Jr. and Graham, J. (1951). On the mass culture of algae. *Plant Physiology* **26**, 539-48.

Nakamura, H. (1963). "Studies on the Ecosystem of *Chlorella*" Tokyo: University of Tokyo Press.

Oswald, W.J. (1973). Productivity of algae in sewage disposal. *Solar Energy* **15**, 107-17.

Pirt, S.J. (1965). The maintenance energy of bacteria in growing cultures. *Proceedings of the Royal Society of London* B **163**, 224-31.

Pirt, S.J. (1975). "Principles of Microbe and Cell Cultivation" Oxford: Blackwell Scientific Publications.

Pirt, S.J., Lee, Y-K., Richmond, A. and Watt Pirt, M. (1980). The photosynthetic efficiency of *Chlorella* biomass growth with reference to solar energy utilization. *Journal of Chemical Technology and Biotechnology* **30**, 25-34.

Pratt, R. (1944). Studies on *Chlorella vulgaris*. IX. Influence on growth of *Chlorella* of continuous removal of chlorellin from the culture solution. *American Journal of Botany* **31**, 418-21.

Pringsheim, E.G. (1946). "Pure Cultures of Algae" Cambridge University Press.

Richmond, A. and Preiss, K. (1980). The biotechnology of alga culture. *Interdisciplinary Science Review* **5**, 60-70.

Sanderson, J.E., Wise, D.L. and Augenstein, D.C. (1978). Organic chemicals and liquid fuels from algal biomass. In "Biotechnology and Bioengineering symposium 8: Biotechnology in Energy Production and Conservation" (ed. C.D. Scott), pp.131-151. New York: John Wiley and Sons Inc.

Sitz, T.O. and Schmidt, R.R. (1973). Purification of *Synechococcus lividus* by equilibrium centrifugation and its synchronization by differential centrifugation. *Journal of Bacteriology* **115**, 43-6.

Sorokin, C. (1959). Tabular comparative data for the low and high temperature strains of *Chlorella*. *Nature*, London **184**, 613-4.

Sorokin, C. and Myers, J. (1953). A high temperature strain of *Chlorella*. *Science*, New York **117**, 330-1.

Stewart, J.R. and Brown, R.M. Jr. (1970). Killing of green and blue-green algae by a non-fruiting myxobacterium *Cytophaga* N-5.

Bacteriological Proceedings p.18

Stewart, W.D.P. and Pearson, H.W. (1970). Effects of aerobic and anaerobic conditions on growth and metabolism of blue-green algae. *Proceedings of the Royal Society of London* B **175**, 293-311.

Taub, F.B. (1969). A biological model of a fresh water community: Agnotobiotic ecosystems. *Limnology and Oceanography* **14**, 136-42.

Thomas, J.P. (1971). Release of dissolved organic matter from natural populations of marine phytoplankton. *Marine Biology* **11**, 311-23.

Vaara, T., Vaara, M. and Niemelä, S. (1979). Two improved methods for obtaining axenic cultures of *Cyanobacteria*. *Applied and Environmental Microbiology* **38**, 1011-4.

Vance, B.D. (1966). Sensitivity of *Microsystis aeruginosa* and other blue-green algae and associated bacteria to selected antibiotics. *Journal of Phycology* **2**, 125-8.

Vela, G.R. and Guerra, C.N. (1966). On the nature of mixed cultures of *C. pyrenoidosa*. *Journal of General Microbiology* **42**, 123-31.

Vincent, W.A. (1971). Algae and lithotrophic bacteria as food sources. In "Microbes and Biological Productivity" (eds. D.E. Hughes and A.H. Rose), pp.47-76. 21st Symposium of the Society for General Microbiology. Cambridge University Press.

Wagner, F., Kleeman, T. and Zahn, W. (1969). Microbial transformation of hydrocarbons, II. Growth constants and cell composition of microbial cells derived from n-alkanes. *Biotechnology Bioengineering* **11**, 393-408.

Ward, C.H., Moyer, J.E. and Vela, G.R. (1964). Studies on bacteria associated with *Chlorella pyrenoidosa* TX71105 in mass culture. *Developments in Industrial Microbiology* **6**, 213-22.

Weinmann, V.E., Walane, P.L. and Trainor, F.R. (1964). A new technique for obtaining axenic cultures of alga. *Canadian Journal of Botany* **42**, 958-9.

Whitton, B.A. (1965). Extracellular products of blue-green algae. *Journal of General Microbiology* **40**, 1-11.

Whitton, B.A. (1967). Studies on the toxicity of polymyxin B to blue-green algae. *Canadian Journal of Microbiology* **13**, 987-93.

Wieringa, K.T. (1968). A new method for obtaining bacteria-free cultures of blue-green algae. *Antonie van Leeuwenhoek. Journal of Microbiology and Serology* **34**, 54-6.

Williams, F.W. (1971). Dynamics of microbial population. In "Systems Analysis and Stimulation in Ecology 1" (ed. B.C. Patten), pp.198-267. New York: Academic Press.

SUBJECT INDEX

Acetate degradation, 71
Acetobacterium woodii, 74
Acetylation, effect on floculation, 132,133
Acid production, in yogurt, 100
Aerobacter, 156
  aerogenes, 43
Aggregation, 121
Alcaligenes, 158,159,165
Algal-bacterial interactions, 156
Algal cultivation plants, 164
Algal ponds, 86
Ammonium oxidation, 7
Anabena, 155
  cylindrica, 166
Anaerobic digestion, 53
Anaerobic filter, 57
Anaerobic nitrogen assimilation, 66
Anaerovibrio lipolytica, 68
Animal feed, production using yeast/*lactobacillus*, 144
Aspergillus oryzae, 137, 143
Asterionella japonica, 157
Axenic algal culture, 153

Bacterial contamination
  of algal cultures, 151
  of beer worts, 122
Bacteroides, 66,67
  ruminicola, 64
Bacillus, 85,156
  polymyxa, 166
Batch digestion, 58
Beer spoilage, 121

Benzoic acid, 14
Biochemical oxygen demand, 86,89
Biomass, separation of, 91
Biotreatment, 83
Brettanomyces, 139
Bulking of biomass, 85,92

Calcium, in cosedimentation, 130
Candida krusei, 143
Carrying capacity, 26
Casein hydrosylate, stimulation by, 109
Caulobacter, 156
Cellulolytic bacteria, 63
Cell wall composition, in yeasts, 130
Chemical waste waters, 82
Chlorella, 153,156,166
  pyrenoidosa, 156,157
  vulgaris, 154
4-Chlorocatechol, 93
Chlorophyta, 153
Ciliate protozoa, 60
Clostridium, 64,65,67
  aceticum, 73
  butyricum, 66
  cellobioparus, 69
  thermocellum, 63,73
Coke ovens, 90
Collected wastes, 83
Commensal relationships, 4,6,34
Competition, 27,31,33
Continuous enrichment, 152,155
Continuous yogurt manufacture, 155
Cosedimentation assay, 126,127,128
Cyanobacteria, 155

SUBJECT INDEX

Cyanophyta, 153

Dairy industry, 99
Dalapon, 14,15
Designated cellulose, 62
Desulfotomaculum, 67
Desulfovibrio, 67
Diazinon, 16
Dictyostelium discoideum, 43
Digester feedstocks, 54
Digester systems, 54
Diversity index, 45
Domestic wastes, 82

Eigenvalues, 47
Energy balance of algal cultures, 160
Energy farms, 57,59
Escherichia coli, 122
Eugleminae, 153

Feedback digester, 57
Fertilizer, 59
Fibre breakdown, 72
Filamentation, 85,92
Flavobacterium, 156,159,165
Floating roof digesters, 62
Floc formation, 86,91,92,122
Formic acid, stimulation of growth by, 103,111
Free amino acids, in milk, 101

Gause's principle of competitive exclusion, 32
Geuze beer, 139
Ginger beer, 140

Hafnia protea, 127,130,131,132, 133
Hydrogen utilizing methanogens, 70
Hyphomicrobium, 5

Industrial wastes, 81

Kefir, 145
Koumiss, 145,146

Kvass, 139

Lactic acid, inhibition of growth rate, 108,111
Lactobacillus, 122,123,145
 acidophilus, 147
 brevis, 123,124,125,126
 bulgaricus, 99,100,101,102,103, 104,109,110,111,112,113,114, 115,146,147
 casei, 35
 caucasicus, 146
 sanfrancisco, 142
Lambic beer, 139
Leuconostoc, 146
Liapounov analysis, 42
Light-dark cycle, in algal biomass production, 165
Light limited culture, 160,162
Lignin, 67
Lipid feed stocks, 68
Logistic equation, 26
Lotka-Volterra equations, 26,37, 40,41,42,44,45
Lyngbya-Prioridium-Plectonema group, 155

Maintenance, 35
Methane-utilizing community, 5
Methanobacillus omelianskii, 14, 69
Methanobacterium
 formicicum, 68,70
 ruminatium, 69,71
 thermoautotrophicum, 73
Methanol inhibition, 17
Methanosarcina, 70
 barkerii, 71,73,74
Michaelis-Menten, 86
Micrococcus, 156
Microcystis
 aeruginosa, 156
 pulverea, 157
Mima, 156
Monod kinetics, 26,28,30,34,40,90
Multispecies systems, 44
Mutualistic relationships, 4,6,34
Myxophyceae, 155

Nocardia, 84
Non-viable bacteria, role of, 83

SUBJECT INDEX

Nostoc, 157
Nitrification, 33,90
Nitriloacetate, 8

Oscillatoria, 155
Oxidative capacity, of bio-treaters, 87
Oxygen uptake, 87

Passenger bacteria, 60
Pepsin, effect on flocculation, 131
Phenol degradation, 86,90
Phenoxyacetate assimilation, 87
pH-stat, 107
Phytoplankton, 151,156,160
Pichia saitoi, 143
Picolinic acid, 19
Picramic acid, 90
Pig waste, 56,61,64,70
Plug flow reactor, 86,115
Polymeric carbohydrate breakdown, 62
Potential oxidative capacity, (Pox), 87,88,89
Predation and parasitism, 36
Product inhibition, 106
1,2,propanediol assimilation, 87
Protocooperation, 8,104
Pseudoanabena, 155
Pseudomonas, 156,158
 aeruginosa, 90
 citronellis, 93
 putida, 19,84

Respiratory quotient, 88
Routh-Hurwitz criteria, 49
Rumen, 58,67

Saccharomyces, 139
 cerevisae, 35,122,123,124,126, 129,130,132,133,143
 inusitatus, 142
 kefir, 146
 rouxii, 143
Saké, 122,137
Scenedesmus, 166
Serratia, 156,158,159,165
Sewage, 55,57,91

Skeletonema costatum, 157
Sludge, 55,59,82,91,92,93
Sludge-blanket digester, 57
Solar energy, 151
Sorghum beer, 139
Sourdough, 142
Souring, of milk, 102
"Sourmash bourbon" whiskey, 140
Soysauce, 143
Stability, 44,163
Staphylococcus, 156
Starch degradation, 65
Streptococcus
 lactis, 67
 thermophilus, 99,100,102,103, 104,106,108,109,110,111,112, 113,114
Substrate stimulated respiration, 87
Symbiosis, 25
Synergism, 8,65
Syntrophism, 4
Syntrophobacter wolinii, 68

Tetrahymena piriformis, 43
Thiobacillus thioparus, 90
Torulopsis
 holmii, 143
 kefir, 146
Transients in substrate concentrations, 87
Trichosporon cutaneum, 144
Two-phase digester, 69

Urea, consumption by algae, 163

Verhulst-Pearl equation, 26
Volatile fatty acids from milk, 103,104

Wall growth, 112
Wort sugars, influence on co-sedimentation, 126

Xanthomonas, 157

Yeast-bacterium aggregations, 122
Yeast strain, effect on bacterial numbers in beer, 123